Dr Willie McCarney was a university lecturer, training teachers to work with disaffected youth. He was also a lay magistrate in Belfast's Youth Court and Family Proceedings Court for 34 years. He is a past president of the International Association of Youth and Family Judges and Magistrates. He travelled the world, on behalf of the United Nations Development Programme, UNICEF and the Council of Europe, training judges in the use of international instruments concerning the rights of the child. In his travels, he gained valuable insights into the various countries and the lives of the people who lived there. This book recounts the experiences and insights garnered over 30 years of travel.

To my sister Una and my son Liam for their support in all that I did. To all the men and women around the world who carry on a never-ending struggle defending the rights of the child.

Willie McCarney

TRAVEL WITH A GAVEL

AUSTIN MACAULEY PUBLISHERS™

LONDON * CAMBRIDGE * NEW YORK * SHARJAH

A CIP catalogue record for this title is available from the British Library.

ISBN 9781398436749 (Paperback)
ISBN 9781398436756 (ePub e-book)

www.austinmacauley.com

First Published 2022
Austin Macauley Publishers Ltd®
1 Canada Square
Canary Wharf
London
E14 5AA

I am indebted to my sister Una and my son Liam. Without their influence this book would not have been written. They convinced me that it was a necessary supplement to Big Boys Don't Cry. They commented on the early drafts, drawing my attention to things I had left out; they were meticulous in the proofreading. I would also like to thank my good friend Tom McGonigle for providing valuable feedback on earlier drafts.

Table of Contents

Preface   11

Introduction   13

Argentina   36

Australia   41

Brazil   52

Canada   63

Chile   69

China   71

Cuba   83

Ecuador   86

France   103

India   108

Iran   112

Italy   116

Japan   122

Jordan   130

Kosovo   132

Liberia   135

Mexico   140

Myanmar or Burma?   143

Nepal 155

Northern Ireland 159

Palestine 165

Peru 167

Poland 175

Russia 177

Scotland 182

South Africa 184

Sweden 203

Switzerland 204

Tajikistan 209

Tanzania 212

Thailand 224

Tunisia 227

Turkmenistan 229

UK 233

USA 235

Travel Broadens Our Mind and Widens Our Horizons
Does It Have to Be Global Travel? 254

A World Apart: A Peek Behind the Curtain of Ignorance 261

Conclusion 276

Countries I Have Been To 281

Timeline 282

Preface

Let me begin by saying that this book it is not a travel guide. If you spend the winter months yearning for sun and sand, luxurious villas and five-star restaurants, you are well provided for with conventional travel books. My book has little to offer you.

"Travel with a Gavel" complements *"Big Boys don't Cry.*[1]*"* It builds on, and adds detail in relation to my work-related travels. Here, I am recounting my unique experiences of the countries I have visited.

I have been one of the luckiest men alive, to have seen so much of the world. I have visited a lot of interesting places. Many were not on the tourist trail at the time of my visit, although they might well be now. But then, I was not there as a tourist. I was there by invitation to train judges in the use of international instruments concerning the rights of the child.

My travels enabled me to gain insights into the countries visited and the inhabitants which would not be available to the average tourist. I want to share these insights with you. If you would like to see the countries I report on through a different lens when you visit then perhaps this book is for you.

The travel chapters draw on diary records, written contemporaneously by my son or my sister, if one or other was travelling with me, or in emails I sent home to let them know where I was and what I was doing. I have noted the year(s) I visited each country in their chapter title to help you contextualise the report in terms of prevailing local and world politics.

I have arranged those countries I write about in alphabetical order, rather than in date order, so that you can easily locate the ones you are most interested in.

---

[1] Big Boys don't Cry" An Autobiography by Dr Willie McCarney, published by Matador, 2015. Available as an ebook.

I have tried to include sufficient information about my background so that the reader will understand where I am coming from without reading my autobiography. But, clearly, compressing the first 14 chapters (167 pages) of the autobiography into 14 pages for the introduction to this travelogue means that a lot of detail has been left out. Anyone wanting more detail should have no difficulty finding what they are looking for by simply checking the chapter headings in *Big Boys don't Cry*: Maesgeirchen – Chapter 1; Marketcross – Chapter 2; Glen Upper – Chapter 3: etc.

Introduction

I was a most unlikely traveller. I grew up in the townland of Loughmacrory in Co Tyrone, Northern Ireland. I loved the countryside and there was nothing I liked more than wandering across the fields with my dog. I loved the peace and the solitude. I loved to listen to the birds singing and search for their nests in the Spring; watch the rabbits at play; keep an eye out for hares and hope to catch a glimpse of a fox, badger or stoat.

I had no great ambitions to travel other than to visit friends and family within a two or three-mile radius or the occasional trip to Carrickmore (the local village) to buy new clothes. From age 11, I had to take the bus each day to the Christian Brothers' Grammar School (CBS), 10 miles away in Omagh. Apart from that there was an annual (one-day) bus trip to Bundoran, a small seaside town in County Donegal. That was more than enough travelling for me.

St. Augustine said: "The world is a book, and those who do not travel read only a page." At age 19, I had never been to Belfast or Dublin, and didn't feel I had missed anything. 62 years later, as I sit down to write this, I realise that I have visited 75 countries in the intervening years.

How had I morphed from someone with little interest in travel to someone who was ready to fly off to anywhere in the world at the drop of a hat? A butterfly only emerges from the chrysalis if the early stages of development have been completed successfully. Could the same be true of globe-trotters? Are the wanderlust seeds sown in the formative years?

I was born in Maesgeirchen, near Bangor in North Wales, in 1938. My parents, and three older brothers came from Co Tyrone to live there when my father found work on Menai Bridge. I don't remember much of the early years. The outbreak of WWII went largely unnoticed. There was nothing in Maesgeirchen that would attract German bombers when war broke out. It should have been one of the safest places in Britain. And yet that is not how it turned out. Some aspects of the war are engraved in my memory.

All the men in the community, including my father and uncles, joined the Home Guard, although I do not have any clear memories of that. I do have memories of the blackout. A surprising number of German bombers found their way across country and flew overhead. There were conflicting views as to what they were doing over North Wales but the most likely explanation is that they were simply lost – unsure where they were because of the blackout in the cities across the country. Once they reached the Welsh coast, they knew they had gone too far and turned back. When we heard the bombers coming, we huddled with our mother under the stairs. On occasion the pilots, presumably low in fuel, would ditch their cargo of bombs to enable them to get back to Germany. The most frightening event for me was one night when a bomb fell near our house. My brother Paddy and I were in bed upstairs when the explosion shattered the window and we were covered in glass. I don't know why we weren't under the stairs and can only assume that either the air raid siren had not gone off or that everyone thought that the bombers were already heading back in the direction of London and had been taken by surprise by this one.

The memory of that night was to stay in my subconscious for many years. As I grew up, I couldn't understand why I felt a sense of dread when I heard the droning of a plane in the distance. I wasn't afraid of planes; I wasn't even afraid of flying. Why should the noise of a plane in the distance get my heart beating faster? It was a long time before I put two and two together. The sound of a plane in the distance awakened memories in my subconscious of the nights we hid under the stairs in Maesgeirchen, listening as those bombers got closer and praying that they would soon turn and fly off again without dropping any of their bombs.

An image which sticks in my mind is of a parachute descending from the sky. I have no idea what happened. I hadn't heard any planes and hadn't heard any anti-aircraft fire. And yet, there it was, drifting by. I had no idea where it came from or where it went to. My cousin Bridget tells me that she recalls her dad (my Uncle Ted) mentioning that a German pilot had been captured. So, it is likely that the apparition I see drifting silently past my mind's eye was that pilot. But whatever it was, I was too young to understand the significance of what was happening. I just gazed in wonderment at this strange apparition before turning my attention to something else. Later, when I saw my oldest brother, Joseph, with some silk parachute cord which he was treating as a prized object I didn't tie the two incidents together. I just couldn't understand why a piece of white cord was so precious.

A narrow-gauge railway transporting slate from Penrhyn Quarries in Bethesda to Port Penrhyn at Bangor ran through the woods close to our house. The steam locomotive was a great source of interest to the local children who played in the woods and watched for the train coming. The first adventure I remember was attempting to follow my three older brothers to see the train. They ran on and left me behind. I fell and hit my head on a rock. I recovered consciousness as Joseph carried me back home.

In 1943 our family, now seven strong (my sister, Mary, was born in 1940), headed back to Co Tyrone. We lived with Aunt Roseanne and Uncle Henry on their farm in Marketcross for a while. Henry did not farm the land himself. He rented it out to local farmers. I hadn't enrolled in the local school yet and there was nothing I liked better than to see Uncle Peter arrive with his horse and cart on the way to harvesting the hay in the meadows out near Lough Fingrean. He would always let me climb up on the cart and go with him.

Just on the boundary of the furthest meadow was a spring well with the sweetest water I have ever tasted. The well was called "Dochaile". The water was crystal-clear and icy cold as it bubbled up from the rocks. A drink from this well was heaven – particularly on a warm summer's day. On the outflow from the spring grew a mass of beautifully tasting watercress which we used to eat in handfuls – excellent for lunch!

One could not imagine anywhere more remote from the war in Europe than Marketcross. True, US army personnel had been posted to various parts of Northern Ireland between 1941 and 1944. Units were based in Omagh, about eight miles from Marketcross, between 1942 and 1944 and in Gortin, about the same distance away, in 1944. A small number were based in Loughmacrory Lodge, home of Sir Hugh Stewart. However, this was not something that intruded into my consciousness at that time.

Then, one day, I heard the roar of a powerful engine and went out to see what appeared to me to be a monster vehicle[2], with a row of monster wheels either

---

[2] I now know that it was probably an M8 Greyhound six-wheel-drive armoured vehicle as these came into service with the US military in late 1943. Few American war vehicles were as important, or so often forgotten, as the M8 Greyhound series. Its "go anywhere, do anything design" was described as "second to none". This, combined with the "can-

side, coming down the lane towards the house. The vehicle stopped on Henry's street. The commander produced a map which he showed to Henry explaining that they intended to proceed past the house in the direction of Lough Fingrean and, presumably, join the Cookstown Road at Creggan. Henry said that they couldn't do that.

To tell American soldiers that they can't do something is like waving a red rag to a bull. They were going to do it anyway. They piled back into their vehicle and roared off following the old cart track towards Henry's meadows. But they didn't get far. A few yards from the house the track narrowed as it passed through a small ravine cut into an outcrop of rock which traversed the lane. The rock had been chiselled away many years previously to allow access to the fields beyond leaving a track just wide enough for a horse and cart. Clumps of heather hung down the sides of this small ravine so that it was not immediately obvious that there was a six-to-eight-foot wall of solid rock on one side and a four-foot wall on the other.

Henry advised the soldiers again that they couldn't get through but they were not taking no for an answer. They didn't seem to appreciate that even a Sherman tank would not have been able to push the mountain to one side. They pressed forward with the engine roaring until the vehicle was stuck tight between the rocks. At this point the vehicle would not even reverse out. Several hours later after a lot of hard labour with picks and spades they managed to extricate their "Greyhound", got it turned around and roared off the way they had come.

I seemed to have had an affinity for wildlife from an early age. Once I was up on the hill looking for a mischievous leprechaun, Uncle Henry had told me about when I stumbled over a hare lying in the heather. The hare made no effort to get up and run away. It just lay there breathing deeply, clearly totally exhausted. I could hear the sound of voices in the distance and the barking of dogs. I guessed that the dogs had been after the hare which was now so tired it couldn't manage another step. It made no effort to escape as I picked it up in my arms. I held it tightly as I scrambled through the heather and over the stone ditches for about half a mile until I was sure that the dogs would have no chance of finding it. When I put it down on the ground it bounded off over the hill without a backward glance.

---

do attitude of the crew" led to it being described as one of the unsung American heroes of World War II.

Looking back now the time spent in Marketcross seems to have been comparatively short. But I am sure it didn't seem that way to my aunt. We had been putting a heavy strain on her hospitality for too long when, eventually, my father found a house to rent not too far away. It was just about half a mile away as the crow flies – down to the bottom of the glen, across a small river and up through the hazel woods. It must have been a great relief to my aunt when the seven members of my family moved out and into our new home in "Glen Upper". We could see Aunt Roseanne's house on the other side of the valley nestled on the side of the hill below Crush Rock. It was such a relief for us, as it must have been for her, that we now had a house to ourselves.

My father had a good job in Wales and we were comfortably well off. Back in Ireland the only job he could get was working for the County Council earning just £3.10s a week – barely enough to keep us off the breadline. To say that we didn't have much money would be an understatement.

Our new home didn't have the amenities we had enjoyed in Wales – there was no running water, no flush toilet and the nearest shop was about a mile away. But the green rolling fields, the tall trees, the hazel woods and the rippling stream flowing through the glen provided opportunities for many new experiences. This was sufficient compensation for me.

Our new landlord, another Henry, told my father that he could have two drills in one of his fields to grow potatoes and vegetables of his choice.

We got hens so that we had our own free-range eggs. I got a "setting" of duck eggs and put them under a broody hen to hatch. Mother hen knew instinctively that chicks should stay away from water and couldn't understand why her "chicks" refused to heed her warnings, made a beeline for the river and refused to come out when she called them. She was a good mother and I watched her standing in the water, her lower feathers soaking wet, so that she could keep an eye on her brood. She reared them successfully. We now had duck eggs to eat also.

We got two goats which meant we had a plentiful supply of fresh milk. Goats are easy to keep and will eat practically anything – grass, gorse, shrubs. I could tell what they had been eating from the flavour of the milk!

Clearly, there was little likelihood of us being dressed in the latest fashion. But, more importantly, there was no chance of us going hungry.

When we came to live in Glen Upper, Henry, still kept some of his cows in a byre at the side of our house. His daughter, Mary Ann, would be over every morning and every evening to tend to the cattle. She was a young woman in her early twenties. She was big and strong and could throw sacks of potatoes onto the tractor with hardly any effort. I used to watch out for her coming so that I could "help" her. She would always give me a piggyback before she headed back home.

I rarely saw Henry's son John. I think he had a job in a local quarry, so he was seldom about during the day. I never saw any of the family at mass on Sunday and asked my father about it. He told me that they didn't come to mass on Sunday because they were not Catholics. "Not being a Catholic" didn't mean a lot to me – apart from not seeing my neighbours on a Sunday morning. They were such nice people. I expected to see them every day and wondered why Sunday was different.

Our landlord told my father that we could cut firewood from the local hazel wood and asked only that he managed the cutting so that no area was left denuded and no gaps were left in fences. We should on no account touch the fairy thorn which stood at Dunnaminfin.

Dunnaminfin stood on a hill a short distance from the house and was believed to be a fairy fort. At the bottom of the hill, just below the fairy fort, was a small area, little bigger than a double grave and completely enclosed by shrubs. Henry insisted that it was a fairy graveyard. It may well have been a grave, perhaps dating back to famine days.

I started school in Wales and had just completed my first year when we came back to Co Tyrone. It was time to continue my schooling. School was in Loughmacrory, some three miles away if we took a shortcut through the meadows, using the steppingstones to get over the river. The river was little more than a gently flowing stream most of the time. But, following heavy rain, it became a raging torrent, swollen by streamlets running down from the sides of the glen. In flood the river could be twenty feet across with flood waters extending well beyond that into the meadows. In full spate it was impassable and we would have to make a long detour which nearly doubled our journey. My father decided to build a bridge across the river – just where the steppingstones were. His first attempt was washed away when an unusually severe flood swept through the meadow. His second attempt was a much more stable construction. It was a proper bridge, complete with handrails, made from hazel branches, which are both strong and supple. We were very glad of the bridge during the years we stayed in Glen Upper.

There were two chimneys on our house, but we never lit a fire in the bedroom, much to the delight of the local jackdaws. The chimney was an excellent location for their nest and they hatched out a brood there each year. We were happy about that too as we were able to climb up on the roof and observe the growing chicks. One year the parents abandoned two chicks. We did not know why but suspected that Henry, our landlord, had shot them. Jackdaws were just pests to him. In any event, we took the jackdaw chicks down and hand-reared them. We called them Jack and Jill. We had no way of knowing whether or not we had got the sex right, but the jackdaws didn't seem to mind. Jill was very tame and would stand on the kitchen floor flapping her wings and squawking until my mum dropped crumbs into her mouth. Jack was always more cautious.

When they got old enough to fly, we set them free. At first, they stayed mainly in the tree in front of the house, coming down when we called them to get fed. Jill was still very tame and responded immediately when she was called. Jack was still more cautious and was always very wary – ready to fly off the moment danger threatened.

As they became more mature, they spent less time in the tree at the house – sometimes they would be missing for several days. But they remembered their names and would come when called if they were within hearing distance. Then Jack stopped coming. We liked to think that he had found himself a partner and had settled down somewhere else but feared that he may have fallen foul of Henry's pest control initiatives.

Not far from our house was what used to be a small lake. When we arrived in Glen Upper it was largely overgrown with reeds and rushes so that it was more of a marshy area than a lake. It was an excellent habitat for frogs and moorhens. In early February, frogs would arrive from all around to mate and spawn. I used to love to hear the incessant "crooning" which went on day and night for a week or so as mating took place. They could be heard distinctly from the house which was several hundred yards away. I found it a comforting sound, perhaps because it was a harbinger of Spring.

It was interesting to watch the frogspawn develop; the tadpoles appear and turn into baby frogs. I was always amazed at how soon the baby frogs would leave the safety of the water and head off into the fields – they were such tiny little things in such a big world. I wondered how many survived and grew to adulthood. It would have been nice to know how many returned the following

Spring to where they had been born in order to play their part in the propagation of their species.

The lake was home to several families of moorhens also. Because it was shallow and largely overgrown it was easy to find the nests in the Spring and watch for the chicks to hatch. I was intrigued to see the chicks take to the water as soon as they came out of the eggs. I was fascinated to see the size of their feet and marvelled at how well they could swim despite having no webbing between their long toes.

I had been introduced to farming life by Uncle Peter when we lived in Marketcross. I am sure that I was more of a hindrance than a help but Peter was fun to be with and I think maybe he enjoyed the company also. Working with Henry (our landlord in Glen Upper) was different. Work meant work to him. I could work well when it came to gathering potatoes but wasn't much use in the corn as I hadn't got the hang of tying the corn into sheaves. Henry grew flax and I found pulling flax a pleasant task. He also let me help with putting the flax into the dam. I enjoyed tramping the flax down while the men put heavy stones on it to make sure it stayed under the water. Taking the flax out again a couple of weeks later was a different story. The water had done its job. The hard outer stem had begun to rot. The smell was abominable. I can still feel the hot summer sun on my back, see the steam rising from the flax as we spread it out to dry and still smell the stench in my nostrils. I was sick all day and couldn't face the thought of eating anything when we broke for lunch or tea. I decided that spreading flax was not for me.

A rocky outcrop at the side of the house meant that there was a gap where the hazel bushes could not grow so that there was a green grassy area which ran down to the river. We used to take the turf barrow, run to get up a little speed and then lie down on it as it careered down the hill. The idea was to try to get up enough speed so that the barrow would hit the riverbank and jump across to the other side. I am not sure that this would ever replace tobogganing as an Olympic sport but we found it great fun even though we ended up in the river more often than on the further bank.

Two sisters lived on the hill on the other side of the river along the path we took when going to church or to school. To me, at that time, they appeared very old but they were probably only in their 50s. They made the most fantastic homemade butter and we were always so pleased when they had some to spare.

They also kept turkeys which were new to me. I was fascinated by their size and by their "gobble, gobble, gobble" when anyone approached.

But of more interest to us, in the context of adventure, was a large duck pond at the side of the house. Memory plays tricks on us as we get older and it always appears that the summers were hotter and the winters colder in our young days than they are today. Looking back now it appears that the duck pond was frozen every winter and we loved skating on it.

Skating for us was not the Torvill and Dean variety. "Dancing on ice" was not something which entered our heads. We had never heard of "skates". Special footwear for us was a good pair of hobnailed boots. The idea was to run on the grass and get up as much speed as possible before stepping onto the ice and attempting to slide further than anyone else. The prize went to the one who could slide the furthest although "prize" is too strong a term – "accolade" maybe, or perhaps just "plaudit". The important thing is that we had great fun.

While the winters might not all have been colder than today, one winter sticks out in my mind – 1947. This was the "Year of the Big Snow – the coldest and harshest winter in living memory". The temperatures rarely rose above freezing point from January to the middle of March. Snow fell on 30 days between January 24 and March 17. The snow which fell in January never got a chance to melt. As snow continued to fall it simply piled higher and higher. Then we had "The Blizzard of February 25". This was the greatest single snowfall on record and lasted for close on fifty consecutive hours. Driven by persistent easterly gales, the snow drifted until every hollow was filled and the countryside took on the appearance of an Artic landscape. All the familiar landmarks were gone. Everywhere was a sea of white.

The world was at a standstill. It wasn't possible to go anywhere. The lanes and roads were covered in snow to a depth of six to eight feet, with drifts up to fifteen feet deep in places. In many cases it wasn't even possible to know where the roads were as the ditches and hedges were buried under the snow. It was fun for us children but it must have been horrific for the elderly and the infirm as food ran out and it was not possible to get to the shops, and for the farmers who could not get access to food for their animals.

I remember my father and all the men in the community were brought together to try to clear the roads to allow lorries carrying supplies from Omagh to get to the local shops. At times it seemed like an impossible task. My father said that some days they worked for hours clearing the snow only to find that, when they turned to come home, the strong winds had blown the snow into the tracks they had just cleared and they had to dig their way back home again. The freezing temperatures solidified the surface and it was to be three weeks before the thaw set in. When supplies did eventually get through from Omagh they came by caterpillar tractor as this was the only type of vehicle which could make it.

But none of this troubled me greatly. I was as happy as an Eskimo in an igloo which myself and my brother Paddy made close to the house. It was so easy to make. There was no need to cut out building blocks. All we needed to do was to make what was effectively a cave in the snow. The snow was frozen so solid that it held its shape as we cut out the snow underneath. It is little wonder that "The Year of the Big Snow" is engraved in my memory for ever.

I was quite fit in those days. Apart from all the usual running about we had a six-mile round trip to school each day and the same trip on Sunday to mass. I didn't always walk. Sometimes I ran. When I wanted to run, I would take my hoop with me (a bicycle wheel, minus the tyre and the spokes). I would use a short stick – preferably one with a slight curve in it, to guide the hoop as I ran. In doing this I didn't notice the distance and I would be at the school "in no time". There I would park my hoop behind a ditch to be picked up on the way home. It is not really the kind of activity which would interest young people today.

In 1948 we moved to a larger house in Loughmacrory. This house would be more convenient for us as it was less than half a mile from the school and the church. I was sad leaving Glen Upper but thought that perhaps the change would provide opportunities for new adventures.

Things did not get off to a good start. My mother had been ill for some time and had lost two children at birth. My brother, Michael, was born in Wales shortly before we came back to Ireland. He lived for just two hours. My sister Ann was born in October 1944, shortly after we moved to Glen Upper. She lived for only three hours. Cause of death in both instances: "Premature Birth". My mother never really recovered after Ann's birth. She died a few months after our move to Loughmacrory. The cause of death was "Arteriosclerosis". She was just 38. I was 10.

With everything that had happened in my short life, it would have been easy to get very depressed. There is one thing however that is proven to be good for lifting the spirits and keeping us healthy both in body and mind, and that is love of nature. While nature cannot cure all of our ills, it can alleviate stress and bring us inner peace. We don't have to be expert birdwatchers or wildlife photographers. It can be as simple as watching a Lapwing pretending to have a broken wing in order to lead a predator away from its nest; spotting the skylark "That from heaven, or near it, Pourest thy full heart"[3] or admiring the breathtaking beauty of a spider's web glistening in the morning dew.

I grew to like Loughmacrory, despite the inauspicious start. I taught myself to swim in the lake. I was very surprised that no one else went swimming there. When word got around that I was swimming in the lake I was told that I shouldn't. I was given no explanation as to why I shouldn't swim there so I carried on. I went swimming most evenings over the summer months. My brothers Paddy and Eddie and our cousin Michael often accompanied me but, if they were not available, I would swim alone.

I seldom went for a walk with my dog that I didn't find something new or exciting – a baby mouse struggling to climb a rock, tumbling back down twice before summiting successfully; two baby long-eared owls, brave enough to venture out of their nest onto a nearby branch but not confident enough to attempt to fly away as I approached; a litter of new-born hedgehogs in an overgrown garden at Sir Hugh Stewart's home; an otter in a local river. I was intrigued by the natural world and how it slipped seamlessly from one season to the next.

Historically, in Ireland, the first day of February was known as "Imbolc". In the old Irish Neolithic language Imbolc means literally "in the belly", referring to the pregnancy of ewes – the cycle of life beginning again. But we knew the first day of February as the Feast of St. Brigid, celebrating the arrival of longer, warmer days. It was the first day of Spring. The natural world awoke after the winter break with a new burst of energy. I watched the buds on the trees begin to swell, the first crocus burst into bloom, the frogs arrive and begin to spawn, the first lamb, the baby rabbits. I listened to the birds singing to establish territory. By March the buds on the trees had burst into leaf; the mad March hares were boxing in the meadows; the birds mating and nesting; the bumblebees and the first butterflies searching for early blossoms.

Farmers were busy tilling the fields. The early potatoes should be sown by March 17. There was corn to sow and new-born lambs to be tended. I was happy at the thought of pocket money as the farmers were always glad of a helping hand. Helping out on local farms from Spring through to the Autumn was an

---

[3] To A Skylark Percy Bysshe Shelley 1792-1822

important source of income. Indeed, it was the only source of income. With my father earning £3.10s a week there was no question of a weekly pocket-money allowance for us children.

I was happy to see the first swallow arrive back from South Africa in April and the cuckoo fly in from West Africa but saddened to hear the honk, honk of the wild geese as they set off, in V formation, on the long flight back to Greenland. Life on the farm continued apace. The turf should be cut by Easter.

May 1 was the first day of Summer. The Cuckoo was in full voice looking for a mate and the swifts, too, were back from Africa. Midsummer's Day really was midsummer. The sun reached its zenith and we had the longest days of the year. Chicks born in the Spring were now as big as their parents but still followed them begging for food. Lambs, calves and foals frolicked about in the fields. But not everyone had time to play. Some birds were rearing their second or third brood. The corncrake called all night long and remained hidden in the meadows all day – not the safest place to be when the mowing starts. In July, the cuckoo was already on the way back to Africa.

August ushered in the Autumn. The corn fields took on a golden hue, the leaves on the trees changed colour and some trees were already shedding leaves; the Swifts were heading back to their wintering areas. The blaeberries were ripe for picking and the blackberries almost so. In September the harvesting began in earnest – cutting the corn, digging the potatoes, bringing home the turf. The hazel trees were heavy with nuts and the apples ripe for picking. For days now the swallows had been congregating on the telephone wires noisily chattering amongst themselves. I wondered if they were arguing about the best time to leave. Were they waiting for a bright starry night? They always seemed to leave at night. I would glance up at the telephone wires one morning and they were gone. Or were they waiting for the wind to help them on their way? The wind always picked up towards the end of September. I would later learn that this wind was associated with the Autumn Equinox. But, growing up, I only knew it as the wind that blew the apples down. The leaves were now tumbling down also and the trees were almost bare. The days were getting shorter but the long nights were brightened by the Harvest Moon, the brightest moon of the year, and the sky was studded with millions of stars (no light pollution in those days). The remaining potatoes were harvested and stored in pits. The corn was threshed, the grain packed in sacks ready to go to the mill and the straw built in stacks beside the hay in a sheltered paddock close to the farmhouse. The remaining turf was brought home from the bog and stored in the sheds. Halloween was a celebration of the successful conclusion of the harvesting.

November brought the onset of Winter. The hedgehogs curled up in bundles of leaves sound asleep under the hedges. The wild geese announce their return, noisily honking overhead. The cows and the horses turn their back to the cold wind, while the sheep shelter behind a ditch. All are watching for the farmer coming with hay. The grass is crisp with frost as I walk with my dog over the fields. Some people bemoan the "dark days before Christmas" but I welcome the long nights, because the Northern Lights[4] put on spectacular displays at this time of year. Before we know it, we have reached the Winter Solstice and the days begin to lengthen again, slowly at first – the old folk used to say: "Every day a cock's stride". As we move into January, I look for early signs of Spring. The snowdrops never let me down. They poke their heads up through the snow and burst into bloom at the first glimpse of the sun to announce that the cycle of life begins again.

I don't know what my future would have been had I stayed in that farming community. Since my father was not a farmer, getting employment, even as a farm labourer, might have been difficult. There wasn't any industry in the area where I might have got employment. I might have been interested in becoming a blacksmith but, even in those days, it was clear that work as a smithy was on the decline.

I didn't have a lot going for me. There were plenty of people who doubted that I had what it takes to succeed. The general consensus was that I should not hope for too much or set my sights too high. But I had two very important allies.

I was lucky enough to have a father who saw the value in education. He wanted his children to have the opportunity he never had. He offered me the chance to embrace education as the way to break free from poverty and build a better future for myself.

I was lucky to have a primary school principal in Master Sheerin who convinced me that I could go all the way.

Both men instilled in me the values of honesty, hard work and a commitment to education. They also taught me not to let others define who I am. If I lived by those values, I could achieve anything I put my mind to. They instilled in me an

_____

[4] aurora borealis.

unshakeable belief that if I worked hard and lived responsibly something better lay just around the corner. The message I learned from them was:

I am the master of my fate: I am the captain of my soul[5].

The third critical factor which shaped my future was the passing of the Education Act of 1947. The Government of the day decided that no young person should be denied access to a grammar school due to the inability of parents to pay the fees. This opened up a new world of opportunity for me which would not have been accessible otherwise. The Act came into force in 1948 and I was one of the first to sit the new transfer test which became better known as "the 11+"[6].

When Master Sheerin announced that my cousin Peter and I had passed the transfer test and would be going to a Grammar School in Omagh (the CBS), the older boys[7] questioned the wisdom of our decision. It was a decision I never regretted. Transformation and change are part of nature and the seasons constantly remind us that these occur naturally. We, too, must go through all phases in order to grow and although we might resist change, it will happen regardless. I appeared to be getting more than my share but I had learned that if I was in rhythm with nature, I could flow more freely with the currents of life. The currents of my life were carrying me towards an as yet unknown destination.

I have to admit that the transfer to the CBS was difficult – doubly difficult. It was difficult both from a financial perspective and from an academic perspective.

Financial difficulties were ongoing. The '47 Act ensured that no young person would be denied access to a grammar school due to the inability of parents to pay the fees. But fees were not the only expense. My father had gone back to work in England, where the wages were so much better, but he was still under considerable strain trying to cover the cost of bus fares to Omagh for three of us, and in paying Paddy's fees for the first year. Paddy had been too old to sit the

---

[5] *Invictus* – W E Henley

[6] The 11+ is a selective entrance examination for secondary school, used by both state-funded grammar schools and many private schools in England and N Ireland to identify the most academically-able children. The exam is taken towards the end of Year 5 or beginning of Year 6 of primary school. The children will be more than 11 years old but less than 12 – hence the name "11+".

[7] Prior to the 1947 Act, children not transferring to grammar school or technical college stayed in their primary school to age 14.

11+ but was told that, if he passed the 13+ at the end of year 1, he would be given a scholarship for the CBS. Eddie had opted to attend Omagh Technical College. Our father also provided us with 3d per day to allow us to buy a hot cup of tea to have with our packed lunch. There was no other pocket money but there were many other things where money was required. We didn't have any. This meant that we simply opted out of all of the social activities. I was a good singer but claimed that I couldn't sing to avoid being selected for the choir. I was a good sprinter but avoided participating in the sports on sports day because the winners would be selected to compete in district and Provincial competitions. I was a good footballer but usually sat out the football training because I didn't have the gear.

There was a range of difficulties on the academic side. We had to leave home about 7 am in the morning and cycle three miles to the nearest bus stop to catch the bus to Omagh. We had to hang around for an hour after school before there was a bus to take us home again. We would be home about 6 pm. We then had to prepare our own dinner. We would eventually get down to homework about 7:30 pm with the aid of a single wick oil lamp. I had no reference books or other sources of information. Fellow pupils would talk about getting information from encyclopaedias and the like. I had to depend on whatever textbooks were provided.

Luckily, during my primary school days, Master Sheerin had made education fun. I couldn't wait to find out what was coming next, what new topic was to be learned, new skill to be mastered. My mind was not focussed on learning to pass exams. I just loved learning.

My father had a saying which I took to heart – "If a thing is worth doing it is worth doing well". His words became a guiding principle for me throughout my life. I put my heart and soul into everything I did. I threw myself into my studies and did well in the Junior Certificate Exams which, in those days, were taken at the end of year three. At the end of year five we had the Senior Certificate exams which identified those deemed capable of progressing to Advanced Level courses with a view to selection for tertiary education.

There was no careers guidance in the CBS. It was difficult to know what path to follow for a future career. There wasn't really a wide choice of professions for young Catholics in those days since Catholics had few opportunities for advancement in the majority of professions. I saw my choice as being restricted to becoming a priest, a teacher or a doctor. I opted for teaching. I chose Irish,

Geography and Maths as my three A-Level subjects. I still remember the excitement of researching for a project on the Irrawaddy River Delta and one on the Pygmies in the Ituri Forest for my geography course. I didn't expect that I would ever *use* the information. I never dreamt that one day I would cruise up the Irrawaddy or visit countries in Africa. I just rejoiced in the learning. No thoughts of future travel here. I saw my future as firmly rooted in Ireland. I planned to study Irish at Queens University and then to do the Post Graduate Certificate in Education (PGCE) to qualify as a teacher.

When my results came out, I believed my grades were good enough for a university scholarship. I can't remember the exact number of points required to merit being offered the university route. My points score was well above that so I was disappointed when offered a scholarship to St Joseph's Teacher Training College, but not to Queen's University. I decided to ask "why?" I stated my case to the man in charge of handing out the scholarships in the Omagh Council Office. He listened to what I had to say and then asked if I knew how many university scholarships, he had given out to the CBS that year. I said, "no" to which he replied, "13." He then asked whether I knew how many scholarships he had given out to the Protestant grammar school. Again, I said, "no" and he replied, "5." He then went on to say, "If you think I am going to give any more university scholarships to the CBS you have another think coming. Count yourself lucky to be getting a scholarship to St Joseph's."

It was the first time I had personally encountered discrimination. I had learned from my father the importance of meeting discrimination with dignity and discipline. The lessons he taught me were summed up eloquently by Martin Luther King on the steps of the Lincoln Memorial in Washington DC some six years later (August 28, 1963):

In the process of gaining our rightful place we must not be guilty of wrongful deeds. Let us not seek to satisfy our thirst for [justice] by drinking from the cup of bitterness and hatred.

I was not aware of any way I could complain about the attitude of the man dispensing scholarships in Omagh Council Office, or appeal against his decision. There was no point in "wallowing in the valley of despair". I decided that I must move on and "hew out of the mountain of despair a stone of hope".

Over the years my horizons had expanded to Carrickmore, Omagh and Bundoran. Bundoran introduced a new concept – the border. Up to this point a border was, for me, just a line on a map separating one country from another. This border was for real, with check points, traffic queues and politics. The check points were necessary because the border separated Northern Ireland (politically part of the UK) from the Irish Republic. It was, in effect, an international crossing point and customs officials had the right to search vehicles and occupants in case they were carrying smuggled goods – hence the traffic queues. This particular border has been hotly disputed since it was set up in 1921 – sometimes more hotly than others. More recently, it became a major bone of contention in the Brexit negotiations. It does what all borders do. It makes us think long and hard about who and what we are, what a country is and why it matters. Unfortunately, the inhabitants of Northern Ireland came to different conclusions so that our border not only separates Northern Ireland from the Irish Republic, it separates the two communities.

"Travel is fatal to prejudice, bigotry, and narrow-mindedness, and many of our people need it sorely on these accounts. Broad, wholesome, charitable views of men and things cannot be acquired by vegetating in one little corner of the earth all one's lifetime."

Mark Twain, in *The Innocents Abroad/Roughing It.*

But one little corner of the earth was all I wanted. When I left Loughmacrory at age 19 to train as a teacher in Belfast foreign travel was far from my mind. Rather, I wondered whether I would be lucky enough to get a teaching post in my old primary school when I graduated.

Teacher Training in those days was very different from today. The course in St Joseph's, more commonly known as Trench House, was very much a practical course where the focus was on learning how to teach rather than on academic advancement. Because it was a "practical" course it was not given degree status. So, whether we did a three-year course for primary school teaching or a four-year course for secondary teaching the qualification was a Teaching Certificate. But to my mind it was a much superior course to the one which is followed nowadays. For us the focus was on the art of teaching. Today's teachers are awarded a degree in English, Maths, Geography or whatever – the academic takes precedence over the practical. The current problem with indiscipline in schools is largely the result of teachers graduating who are expert in their chosen subject but weak in the area of teaching skills.

Today's students select a main subject and a second subject. In our first year we covered thirteen different subjects – Art, English, Geography, History, Irish, Maths, Music, PE, Religious Education, Rural Science, Science, Speech & Drama and Woodwork. This list does not include the theory of Education which underpinned everything we did throughout our first three years.

In our second and third year we dropped Irish and Speech & Drama. This meant we still had eleven subjects (plus Education). So, by the end of our third year we had learned how to teach any subject which was likely to be taught in primary school. We were all qualified primary school teachers. Anyone who wished to become qualified to teach in secondary or grammar schools would then opt for a fourth year where they had the opportunity to specialise in a particular subject.

By the end of year three I had decided that I would teach in secondary schools where I could work with under-achieving young people. I loved sport and already had some experience of working informally with young people in Loughmacrory teaching football, handball and Irish dancing. I chose to specialise in physical education believing that I could use sport as a way of getting close to disaffected youth.

I enjoyed my four years in Trench House where I found the majority of the staff very good at their job, some exceptionally good and one or two brilliant. However, I found myself at odds with the culture of education in the college. It appeared to me to be rooted in the work of John Dewey[8]. He argued that education must be primarily geared to the outcome of successful control of one's economic and cultural environment. This conveyer-belt image of education has its origins in the pragmatically orientated, industrial revolution. To be fair, this approach to teaching is very tempting, precisely because it needs little investment of oneself in the process. It considers that, as an educator, one has only to impart specialised knowledge (gained through hard study and experience) to a group of learners. The learners should passively absorb this knowledge.

I developed my own philosophy of education, rooted in the classical approach. Guided by the work of Plato, Aristotle and the ancient Greeks, and the

---

[8] John Dewey was an American philosopher, psychologist, and educational reformer whose ideas have been influential in education and social reform. He is regarded as one of the most prominent American scholars in the first half of the twentieth century.

later teachings of St Augustine, my philosophy is firmly grounded in my own experience.

I saw my mission as the awakening of the dormant potential within each individual child I taught. I believe that it is the role of the teacher to identify skills which the children already have and help them develop those skills to the maximum. Clearly, not every child has an equal talent or an equal ability or an equal motivation, but they all have an equal right to develop what talents they have in order to make something of themselves. Every child should be given a fair chance to be educated to the limit of his/her talents. Success should not be measured solely by the number of children achieving O-Levels and A-Levels. Success should be measured by the extent to which the teachers help *all* children to be the best they can be.

The United Nations Convention on the Rights of the Child (UNCRC) was adopted by the UN General Assembly and opened for signature on November 20, 1989. It came into force on September 2, 1990, after it was ratified by the required number of nations. It was the first legally binding international instrument to incorporate the full range of human rights for children and the most ratified in the history of the UN. 196 countries are party to it, including every member of the United Nations, except the United States. It has a total of 54 articles. No article is regarded as more important than any other. But there are four that are seen as special. They're known as the "Guiding Principles" and they help to interpret all the other articles and play a fundamental role in realising all the rights in the Convention for all children. They are: Non-discrimination (Article 2); Best interest of the child (Article 3); Right to life, survival and development (Article 6); Respect for the views of the child (Article 12).

As I studied the UNCRC, I was drawn to the four "Guiding Principles". It was reassuring to note that three of the four could have been drawn directly from my philosophy of education (Article 6 did not fall within the remit of teachers per se). I may have struck a discordant note with my theory in Trench House but I was clearly in harmony with thinking on the international scene.

But this is jumping too far ahead. The question we had been trying to answer in 1961 was whether I had been bitten by the travel bug during my four years in Trench House.

Many students head off to England to work during the long summer holidays. The majority of those I spoke to were heading for "Birds Eye[9]" where there is always a demand for seasonal workers to harvest the vegetables. Their aim was to make enough money to pay for a holiday in Spain, Inter-Railing in Europe or whatever, where they would blow the whole lot before returning to their studies in Belfast.

I had a different agenda. I wanted to earn as much money as I could and save enough to cover my expenses for the next academic year. I decided against joining the group going to Birds Eye. At the end of the first year, I headed for Manchester where I got a job as a labourer with John Laing Construction. The work was hard but the pay was good. It got even better when I agreed to take down some scaffolding for them. In those days, there was no special training for scaffolders. I could climb like a monkey and had a good head for heights. That was all they needed to know. When I announced that I was returning to Belfast to continue my studies the foreman offered me a fulltime job as a scaffolder. I would be earning much more than I would as a teacher. I thanked him for the offer but said I would stick with teaching.

The following year I was in Spooner's Engineering Works in Ilkley, Yorkshire. I was sweeping the floor for the first few days. Then the manager called me into his office and asked if I could read plans. One of his engineers had gone off on long-term sick leave. Would I stand in for him? This was a non-union shop. The work he wanted me to do was straightforward. It was like building a giant Meccano set. I was building the skeleton of a drying machine (for drying grain). The working parts would be fitted by experts further down the line. Things went well and I was back in Spooner's the following year. The manager offered to put me on a fast-track course to qualify as a mechanical engineer. (He knew I was already qualified as a primary school teacher). The money was very tempting. But the urge to teach was stronger still.

At the end of year four I qualified as a secondary school teacher and began to look for a teaching post. The nearest vacancy was in St Colman's Secondary School, Strabane – about 30 miles from Loughmacrory, but still in Co Tyrone. My application was accepted. I began my teaching career in St Colman's in September 1961.

---

[9] **Birds Eye** is an American international brand of frozen foods owned by Conagra Brands in the United States and by Nomad Foods in Europe.

I had chosen to teach in secondary schools so that I could focus on under-achieving boys who were found mainly in the D streams. Too often these boys were dismissed as having little or no talent. Many turned their back on education. Some got into trouble and found themselves before the courts. I became a Lay Magistrate in 1976 believing that the courts could have a positive impact on the lives of these young people rather than a negative one. I wanted to promote the idea of custody being an option of last resort, reserved for those who were a danger to themselves and/or the community. I wanted to ensure that any decision taken was in the best interests of the child. This should always be the guiding principle.

The thought of foreign travel played no part in my decision to become a teacher or to become a lay magistrate. Had I been interested in foreign travel I would not have considered going down either of these two roads. I never imagined that my unconventional views on education and my focus on the Rights of the Child would bring me to so many interesting places all over the world.

The first indication that I would be required to travel was when I was nominated for the post of Honorary Secretary of the Northern Ireland Lay Magistrates' Association (NILMA) at the AGM in 1982. One of my roles as Honorary Secretary was to be the Northern Ireland representative on the Executive Committee of the British Juvenile & Family Courts Society (BJFCS). This meant attending meetings in London four or five times per year.

Then in August 1986 I was asked to represent Northern Ireland at the World Congress of the International Association of Youth & Family Judges & Magistrates (IAYFJM) in Rio de Janeiro, Brazil.

In 1990 I was elected Chair of the BJFCS. Later the same year, I was elected to the General Purposes Committee of the IAYFJM at their World Congress in Turin. The following year I was appointed Editor-in-Chief of their magazine. At the World Congress in Bremen in 1994 I joined the Executive Committee and four years later, in Buenos Aires, was elected Vice President. I was elected President of the IAYFJM at the World Congress in Melbourne, Australia, in 2002.

Each new role saw a ratcheting up of the need to travel until I was off somewhere at least once a month. Once you begin to travel it becomes part of who you are.

Travel broadens the mind and widens our horizons. It gives us the opportunity to see new sights, experience new cultures, meet new people. We meet people who have a different way of life, who live by a different set of rules. If we keep an open mind and avoid judging others, these new experiences will change our perspective on the people we meet and the countries we visit.

In meeting people of many nationalities, I find that it is important to avoid stereotypical impressions. I work on the principle that a stranger is just a friend I haven't got to know yet. In my view, it is possible to relate to *anyone* in the world if you look past the superficial things that separate you. A smile, and a friendly attitude, can break down barriers and help create friendships. Kindness and respect are the key. Once the "stranger" realises that you respect his/her culture and traditions you are well on the way to making a new friend.

Wherever I went, my love of wildlife went with me. I never go anywhere without my binoculars. I am an early riser so that, even when time is pressing, I can go bird watching in the hotel gardens when everyone else is still asleep. My prize sightings include spotting my first Hoopoe in the garden of the Sedona Hotel in Yangon; a flock of Golden Orioles mixed with Black-Hooded Orioles in Chitwan, Nepal; a Toucan flying across the face of the waterfall at Iguasu from the Brazilian side to where I was standing on the Argentinian side; a Superb Blue Wren in the Botanic Gardens in Melbourne; a Blacksmith Plover and Black Oystercatchers in Cape Town; Inca Terns in Peru.

My love of wildlife is evident in all my reports, in some cases making up the bulk of the report. These include the Valdes Peninsula, a UNESCO-registered nature reserve, and the Tierra del Fuego National Park, both in Southern Argentina; the Cloud Forest and the Galapagos Islands in Ecuador; the Inca Trail and Machu Picchu in Peru and the safaris in South Africa.

I don't comment on all the countries visited. Sometimes I was so tied up with work that I saw little beyond the airport and the inside of some office buildings. Occasionally I was left twiddling my thumbs when the local organisers ran into unexpected difficulties and was then expected to pack a 5 or 10-day programme into a couple of days and still wrap it all up neatly by Friday afternoon. Occasionally, very occasionally, the organisers realised that "all work and no play" impacted poorly on the trainer as well as on the "students" and they set time aside for some sightseeing. You may be surprised to find little comment on major cities like London, Paris, Rome, New York – even where visits to a particular city have been frequent. If there is nothing I can add to what is already available in travel books I don't include it here.

It is said that Marco Polo was asked on his deathbed to admit that all the stories of his life and travels were just a bunch of fairy tales. He answered: *I've told you less than a half of what I've really seen out there.*

There is no way I could recount even half of my experiences of 37 years working for NILMA, the BJFCS, the IAYFJM, the UNDP, UNICEF and the Council of Europe in this slim volume. I hope the experiences I have chosen to include are of interest.

This book concludes my globetrotting, finishes it and makes it complete. But it is more than just a record of my impressions. It is a creative transformation, transforming what I have seen, experienced and learned into something new and unique, something that can be shared. I would like to share it with you.

Argentina

(1986, '97,'98, 2004)

The IAYFJM holds a World Congress every four years and countries bid to host it. Argentina's bid to host the 1998 Congress in Buenos Aires was accepted by the Association's Executive Committee. In November 1997, we arrived in Buenos Aires to review progress. I fell in love with the city. The streets were lined with Jacaranda trees – 80 to 100 feet high, absolutely covered in large purple/blue flowers. They made the streets look so beautiful.

Avenida 9 De Julio (July 9 Avenue) runs through the centre of Buenos Aires. The Avenue gets its name from Argentina's Independence, July 9, 1816. It is about 1 kilometre long and is the widest Avenue in the world. It is a dual carriageway with seven traffic lanes going in either direction although, at peak hour, drivers ignore the lanes. I have estimated twelve to fourteen rows of cars, lorries and buses where there should only have been seven. The scale of the street doesn't hit you until you are standing in the middle of it. I felt no urge to hire a car and get amongst them.

Two of Argentina's most famous monuments, the Obelisk and the Teatro Colón (Opera House), are on Avenida 9 De Julio.

The Obelisk is located in the Plaza de la República in the intersection of 9 de Julio and Corrientes. It was erected in 1936 to commemorate the quadricentennial of the foundation of the city. It stands 67.5 metres (221 feet) tall.

The Teatro Colón is considered one of the ten best opera houses in the world and is acoustically considered to be amongst the five best.

I found Buenos Aires a very relaxed city and was completely at ease walking around its streets. It struck me more as a European city than a South American one. Indeed, one of its nicknames is "Paris of the South". Buenos Aires is also known as the Tango Capital of the world and I saw couples dancing the Tango in the streets.

I have been back to Argentina on six or seven occasions, attending conferences in La Plata and Mendoza, as well as Buenos Aires, and have always enjoyed my time there. But, on this first visit, I didn't stay as long in Buenos Aires as I had expected. The Chair of the Local Organising Committee was in touch to advise us to pick up tickets for a flight to Iguasu – a small town in a beautiful setting on the Argentinian side of Iguasu Falls. We had been booked into the hotel there and could relax by the Falls when we had concluded the business of our meeting. They couldn't have made a better choice, as far as I was concerned. I won't go into details about the Falls here as I cover that in the report from Brazil.

The meeting didn't go as smoothly as we had hoped. The Chair, a big, domineering man brushed aside all questions, responding simply that everything was in order. We were given lots of assurances but very little detail about the organisation of the Congress.

The Congress of '98 was a nightmare, particularly for me. The American delegation had a major row with the Chair – so bad that it almost became a diplomatic incident. I spent a lot of time acting as mediator in that row. There were complaints coming from every quarter. I attended very few of the presentations during the Congress and none of the workshops because I was always so tied up. Strange then that the Congress was a highlight in my career as a Lay Magistrate. I had been nominated for the position of Vice-President and was duly elected unopposed at the General Assembly.

Meanwhile, with me tied up all day every day, my sister Una, who had come with me, was left to fend largely for herself. And she had been having her own problems. We had stopped off in Rio de Janeiro on the way to Buenos Aires at the invitation of a very close friend, Alyrio, a senior judge in Rio. Alyrio had asked me to give a lecture at their Judicial Training College. Una had taken ill in Rio and we thought she would have to be hospitalised. She was not happy at the prospect so I talked a local doctor into treating her "at home" (namely in her hotel room). The doctor, who called on several occasions, had no idea what was causing the problem but was able to get her back on her feet. When we got back to Belfast a consultant in the Royal Victoria Hospital said that she had had a bad reaction to the malaria tablets and she was off work for three months. But in Rio we only knew she was ill and feeling very weak.

I didn't want us to leave Rio without Una having an opportunity to visit the *Christ the King* statue. So, I hired a taxi to take us there. The taxi wasn't able to take us all the way to the top. We had to climb a number of steps. Una was weaker than I had realised. She collapsed when we got to the statue and I feared I was going to have to call for an ambulance. However, she managed to make it back to the taxi, with a lot of support.

Una had recovered a little by the time we got to Buenos Aires. The then Chair of the Northern Ireland Lay Magistrates' Association was also attending the World Congress and her husband, a GP in Belfast, had accompanied her. I felt more confident knowing that I could call on him for help if need be.

I was happy that Una was well enough to explore some of Buenos Aires while I was tied up at the Congress. She likes to pass the time reading. She found a bookshop which sold some books in English but didn't see much of interest. Then she spotted a book by Agatha Christie and decided to give it a try. She has been an avid fan of Agatha Christie since that time. Perhaps that is appropriate since there was so much drama around the Congress of '98.

When the Congress was over, we tried to forget the unsavoury aspects by taking a holiday in southern Argentina. We headed for the Valdes Peninsula, a UNESCO-registered nature reserve that is home to a large variety of wildlife. We went whale watching at Puerto Madryn where we saw *Southern Right Whales*. We saw a large colony of *elephant seals* and *sea lions* at Punto Norte. We went on to see the noisy colony of *Magellanic penguins* at Punta Tombo. Driving inland we saw *rheas* (a close cousin of the ostrich), *guanacos* (a cousin of the llama) and a wide range of birds. We were lucky enough to spot an *Argentine grey fox* – a pretty rare experience we were told. I always associated Gauchos with cattle but we saw several Gauchos rounding up sheep! We visited the town of Puerto Madryn, the site of the first Welsh landing in Argentina in 1865. There are estimated to be some 50,000 of their descendants still living in the area. We visited the town of *Trelew* where many still speak Welsh and retain the Welsh culture with annual Eisteddfods[10]. We stopped at an excellent tearoom where the waitresses wore traditional Welsh dresses. The coffee and tray bakes were superb.

Finally, we flew south to *Tierra del Fuego*, separated from the mainland by the Magellan Strait. We stayed in *Ushuaia*. Situated on the Beagle Strait,

---

[10] An Eisteddfod is a festival of Welsh literature, music and performance.

Ushuaia is the largest city in Tierra del Fuego and, arguably, the southernmost city in the world. It must also have one of the shortest runways. We held our breath as we saw the water approaching fast. But the plane came to a standstill before we reached it. We expected our accommodation to be rather spartan but found ourselves in a luxury five-star hotel with a magnificent swimming pool. My swimming pants had gone missing along the way so I had to purchase a new pair. Apart from swimming we had a great day's bird watching in Tierra del Fuego's National Park. We had a relaxing few days in Ushuaia – a lovely way to finish off our visit to Argentina.

On the flight back to Belfast I promised Una that, when my travel schedule allowed, we would go back to Brazil so that she could see the best of Rio and visit Iguasu.

I will comment briefly on one of my other visits to Argentina. On returning home from India in 2004, I got an unexpected invitation. The Argentine Association had a dilemma which they hoped I could help them with. Their representative on the IAYFJM Executive Committee had been accused of accepting bribes in return for "not guilty" verdicts and had been sent for trial. Did we have any plans to sack him? He was denying the charges and some of his colleagues believed that he had been framed. I said I wasn't sure I could help but was assured that my advice would be gratefully received. They said they needed a mediator who was a good listener, had no hidden agenda and whose advice would be carefully thought out.

I was travelling a lot and had a full schedule of trips lined up. I decided I could only afford a couple of days in Argentina. I met with the group the evening of the day I arrived and again the following day. I told them that I was not there to tell them what to do. I would help them to explore the problem and formulate serious and precise proposals for dealing with it. I advised that in their hands, more than mine, would rest the final success or failure of the course they chose. The International Association would not "sack" their representative. The Argentine Association would have to decide whether or not they wanted him to continue in that role. Did he still meet the high standards they set for those who represented their country? If the majority felt that, in good conscience they could say "yes" because they believed he had been "set up" they needed to ask some further questions. How could he represent their interests if he was remanded in custody or received a prison sentence? Would they be happy with him representing them while on remand if he were to be released on bail pending an appeal? The group was split on the "guilty" – "not guilty" issue. They homed in on my final question. The appeal process was likely to be long drawn out. They decided unanimously that he could not represent their best interests in the interim. They would nominate a successor. I thanked them for their efforts and said that I believed their decision was in the best interests of both the national and international associations.

The group wanted to do something for me to show their appreciation for coming so far to help them resolve their dilemma. I said that a satisfactory outcome was sufficient reward for my efforts. They kept pressing so I said that, if they could tell me where I might see a condor that would be my reward. With a wingspan of 3.5m (11.5 feet) the condor is the largest flying land bird in the Western Hemisphere. Despite many visits to America, north and south, I had never managed to see one. My flight home did not depart until late the next evening so one of the judges volunteered to take me to "The Valley of the Condors". It had been estimated that up to 200 condors lived there. What better place to go to see them? It is a beautiful spot with spectacular scenery. We parked the car and walked for several miles. The towering walls on either side provided excellent nesting places but there wasn't a condor in sight. We spoke with one of the wardens who told us that the birds took off each morning at sunrise. They could be hundreds of miles away by now. If we could stick around until dusk, we would see them all returning to roost – a wonderful spectacle. Unfortunately, that was not possible as I had a plane to catch. I enjoyed my visit to The Valley of the Condors nonetheless.

Australia

(2002)

Organising the IAYFJM's World Congress is a major undertaking and the four-year interval is just about right to allow the country concerned time to raise sponsorship and get everything in place. There are generally numerous countries interested in hosting the Congress but the majority lose interest once they discover just what is involved. The Secretary General is responsible for identifying the serious contenders and advising the Executive Committee. Our Secretary General had taken ill immediately after the Buenos Aires Congress and 12 months later work on the next congress had not even begun. I was asked to sort the problem out. I whittled the contenders down to two. At an executive committee meeting in Singapore, I submitted bids from South Africa and Australia. The Australians proposed that the Congress be organised jointly with New Zealand. The Executive accepted the Australian bid. The Congress would be in Melbourne from the 26 to 31 October 2002.

We were expecting a lot of the organisers. They had less than two years to get their act together. I promised to offer whatever help and advice I could. I flew to Melbourne for a meeting with the Local Organising Committee on April 12, 2002. Because time was pressing, I decided to take a direct flight by way of Los Angeles. It wasn't a good idea. I arrived totally exhausted, as I can never manage to sleep on the plane. I wanted to lie down and sleep on arrival but I had learned from experience that it is better to try to stay awake until bedtime locally. I took a walk in the Botanic Gardens where the bird life is fantastic. I had no trouble keeping awake. The most amazing bird I saw that day was a Superb Blue Wren which landed right in front of me – literally feet away. It hopped around pecking the ground for several minutes before flying off. Of course, I didn't have my camera – it was still in my suitcase!

I had several meetings with the Local Organising Committee over the next couple of days. There were a number of issues still to be resolved but the organisers were totally confident that everything would come together in time. It did.

The Congress was held in the magnificent Melbourne Convention Centre. It was a pleasant walk along the river to the Centre each morning. The theme of the Congress was exploring ways of forging new links in international and interdisciplinary chains of communication towards better delivery of child protection, family law and juvenile justice. There were excellent speakers drawn from around the world. The workshops were well organised and focussed on the theme. There was an excellent social programme.

The IAYFJM General Assembly was well attended. I was proposed for the position of President and elected unopposed. It was a great honour as a lay judge to be elected as President of an association consisting almost entirely of professional judges some of whom sat in the Supreme Court of their respective countries. I was the first person from Ireland, north or south, to hold this position and only the second English-speaking President in 80 years of the Association's history.

Una had decided to come with me on this trip to be present when I was elected President. I decided against a direct flight on this occasion. We considered a stop-over in Bali but decided against it, mainly because I wanted to get a suit made in Hong Kong. I was also convinced that Una would like the city. As events turned out we were fortunate that we chose Hong Kong. The Bali bombings occurred on October 12, 2002 just five days before we left home. 202 people were killed and a further 240 people were injured.

I love Hong Kong with its "spectacular skyline and harbour, a unique and dynamic blend of old and new, East and West, successfully combining western contemporary style with Chinese tradition and heritage"[11]. Una loved it too. We spent hours roaming through colourful street markets and shopping malls. We took a cruise around Victoria Harbour. We went up to *The Peak*, the highest point on Hong Kong Island. Riding the Peak Tram, the city's historic, funicular railway, is a visual experience in its own right. The view from the top is amazing. When we collected my suit, we were ready for the next leg of our journey.

---

[11] Quote from a "Welcome to Hong Kong" pamphlet.

When Una and I arrived in Rio de Janeiro on the way to the congress in Buenos Aires Una had taken ill. This time it was my turn. I could hardly walk when I arrived in Melbourne. I made an appointment to see a doctor the following morning. She diagnosed a septic toe and gave me some antibiotics. I got immediate relief and the pain was gone completely within a couple of days.

Mindful of how close we came to being caught up in the Bali bombing, the first thing we wanted to do in Melbourne was to go to City Hall to see the floral tributes which had been left on the steps in memory of the 88 Australians who had been killed.

We didn't have a lot of free time in Melbourne since most of my time was taken up with the Congress but we made the best of what time we had. Una and I stopped in the Novotel on Collins Street – within easy walking distance of the Congress – so she was able to walk over with me in the morning and meet me in the evening.

Melbourne is a beautiful city. The tramway system is the largest urban tramway network in the world. We used it extensively. We found lots of excellent restaurants and coffee shops. We took photos of the Victorian Railway Station. But mostly, when we had time, we crossed the footbridge over the Yarra River and headed for the Royal Botanic Gardens. The Gardens are home to more than 50 different species of birds including Australian Wood Duck, Bell Miners, Black Swans, Blackbirds, Eastern Rosella (parrots), Grey-headed Flying Foxes (fruit bats), Indian Mynahs, Magpie Larks, Night Heron, Purple Swamp Hens, Red-Rumped Parrots, Reed Warblers, Sulphur-Crested Cockatoos, Superb Blue Wrens (also known as Fairy Wrens), White-faced Heron and Willie Wagtails. I told Una about the Superb Blue Wren and how one had landed at my feet and hopped around for several minutes. I didn't expect to be so lucky again but suddenly there it was, just like before. It landed at our feet and hopped around. Then we had a bonus when it was joined by its mate. They really are superb. Our other favourite birds were the Sulphur-Crested Cockatoos and the Grey-Headed Flying Foxes, usually simply called Fruit Bats. We were disappointed not to see a Kookaburra.

Before leaving Melbourne, we wanted to take an evening tour to Phillip Island to see the Penguin Parade. We were travelling along in the bus looking out for Kangaroos when suddenly we spotted a Kookaburra sitting on the telephone wires. It was our first sighting but after that we seemed to see them everywhere!

We stopped off at Warrook Cattle Farm where we got close up to Kangaroos. The Kangaroos were very friendly and Una got her photograph taken sitting petting one.

Our next stop was the Koala Conservation Centre where we were able to observe them in the trees but didn't touch them. It was nice to see them in their natural habitat.

But the main reason for our trip was to see the Penguin Parade. It was an amazing spectacle to watch these little penguins as they emerge from the surf and waddle to the safety of their dune burrows. It was possible to see them feed their waiting young with the pilchards and anchovies they had caught.

On our way back the bus driver pointed out Albert Park where the Australian Grand Prix is held. We had seen it so often on TV that we decided to have a walk around it next morning before leaving for Sydney.

The first thing we did in Sydney was to get a three-day pass and hop on a ferry to sail around the harbour and get photographs of the Opera House. It was an amazing feeling just to be there. A visit to Bondi Beach was a must so we headed there next morning for a swim and a stroll. We couldn't believe how *small* it was. It looks so much bigger on TV. We are used to beaches which run for miles back home in Ireland, although I would have to admit that it was a bit warmer in Bondi.

The next "must do" on a visit to Sydney is to climb the Bridge. We were looking forward to what has been described as "The climb of your life". Climbers are provided with protective clothing appropriate to the prevailing weather conditions, and are given an orientation briefing before climbing. During the climb, we were secured to the bridge by a wire lifeline. Each climb begins on the eastern side of the bridge and ascends to the top. At the summit, the group crosses to the western side of the arch for the descent. Each climb is a three-and-a-half-hour experience.

We were a little disappointed with the climb. After all the hype, the talk about levels of fitness required and the wire lifeline, we were expecting more. True the view from the top was spectacular but the climb itself was relatively easy. We thought much of the lead-up was pure razzmatazz, intended to make the climb more memorable.

Next day, Sunday, we took a trip to the Blue Mountains. We stopped off at Featherdale Wildlife Park to see the Koalas and Wallabies. This place differed from the Conservation Centre we visited in that it was possible to hold the Koalas and have photographs taken.

We stopped at the rock formation known as *The Three Sisters,* one of the Blue Mountains' best-known sites, but missed one of our other scheduled stops because of bush fires. There was a haze of smoke hanging over the entire area. We visited the town of Katoomba where we stopped at a fine colonial-style hotel for lunch. We were having no luck spotting wild kangaroos so we stopped for a walk through a park before leaving the mountains. We were on our way back to our minibus when I spotted a small group of kangaroos sleeping in the shade of a tree. One female had a Joey poking his head out of her pouch.

On our way back down the mountain we met a number of fire engines racing up. Presumably the bush fires had broken out again. We stopped a few miles outside Sydney and took a catamaran back to the city. We had dinner in a restaurant beside the Opera House overlooking the harbour.

From Sydney we flew to Cairns, the gateway to Australia's tropical north. We were then taken by coach to our hotel in Port Douglas. Port Douglas is just an hour's drive north of Cairns via a spectacular coastal road with the Rain Forest on one side and the Coral Sea on the other. It is the only place on Earth to have two World Heritage listed sites: the Great Barrier Reef and the Daintree/Cape Tribulation rainforest. We had come to see both.

We struck up an immediate relationship with Bozo, one of the long-term residents in our hotel. He found out very quickly that I liked apples and kicked up a row if we tried to get past him without sharing. He was a Sulphur-Crested Cockatoo and was already well into his thirties.

On our first evening we went for a walk and spotted three large lizards sitting at a bus stop. We joked that they were heading into town on a night out!

We had booked a balloon ride for our first full day in Port Douglas. We were picked up very early in the morning and driven most of the way back to Cairns to board our balloon. 30 minutes seemed long enough when we booked the ride but, on the day, we seemed to have barely reached cruising height when we were on our way back down again. However, we would not have missed it for the world.

Daintree National Park is valued because of its exceptional biodiversity. It contains significant habitat for rare species and prolific birdlife – an estimated 430 bird species. Unfortunately, most of them were foraging elsewhere when we visited.

We took the train up through the rain forest to Kuranda. We spent some time in the village and visited an aviary where we saw lots of local birds. However, it is not the same as seeing them in the wild.

We took the Skyrail back into the Rain Forest where we stopped for a guided tour. We then went to visit an aboriginal site called Tjapukai where some of the locals played Didgeridoos to entertain us.

The next day we went to Daintree, visited Cape Tribulation and went on a crocodile hunt on the Daintree River. We saw a very large female lying on the riverbank, basking in the sun.

Friday, 8 November, we headed for the Great Barrier Reef. We took a catamaran to the outer reef. The journey took about 90 minutes. The water was quite rough so we did not go snorkelling. We went out in a semi-sub instead. We were sitting down in the bottom of the boat – surrounded entirely by glass. We had a great view of the fish and the coral.

We flew south to Brisbane where we spent a couple of days walking and relaxing before flying to Wellington, New Zealand where we had been invited to stay with the Chief District Court Judge.

I had received presents from some of the delegates at the Melbourne Congress and had packed them away carefully at the bottom of my case. I was not expecting to take them out of the case until I got back to Belfast. I couldn't even remember what the presents were when taken aside by security at Wellington airport. A security officer informed me that I had tried to smuggle wooden articles into New Zealand – a criminal offence. He produced a form which I had filled in before landing in which I said I had no items made from wood. He informed me that the penalty was a six-month prison sentence. I said I had to accept that I was guilty as charged but insisted that there was no criminal intent. I explained how I had come by the wooden items and why I had forgotten about them. He eventually accepted that my contrition was genuine and said he would let me off with a caution. If I attempted to bring wooden items into New Zealand again, I would feel the full force of the law. My friend David had a good laugh when we told him why we had kept him waiting so long.

David had prepared a reception fit for a President and had prepared a detailed itinerary. We were taken somewhere different every day – I even had a day in court where I had the opportunity to join one of the judges on the Bench. We were hosted to dinner every night either by David and Catherine or by some of their judicial friends. Catherine loaned us her car so that we could go off on our own. The only thing David hadn't catered for was the weather. We expected the weather to be sunny and warm with the temperature around 17°C. Instead, we had rain and sleet with the temperature hovering around 10°C. This was highly unusual and not what anyone had expected. I didn't have any warm clothes with me and went off to buy a warm jumper. The shops didn't have any since we were now in late spring. The only thing I could find was an All-Blacks rugby sweatshirt. Luckily, I am a rugby fan and admire the All-Blacks.

We flew south to Christchurch. We took a taxi in from the airport. When we gave the driver the address of our hotel, he said we were lucky we had not arrived earlier. The street where our hotel was located had been closed all morning because of a bomb scare. And we thought we had left all that behind us! Our driver didn't know if it was a hoax or a false alarm.

After lunch, we went to see the magnificent Cathedral and have a walk around Cathedral Square. We wandered around the pedestrianised sections of Cashel and High streets commonly known as "City Mall", and generally explored the city centre. Looking at the well-developed residential gardens with their many trees and strolling around Hagley Park we could see why Christchurch has been called *The Garden City*. It had the feel of an English country town. It reminded me in particular of Cambridge with the river flowing through the centre which we later found out was called the Avon. Christchurch is well known for several very traditional schools of the English public-school type such as St Thomas of Canterbury College. We thought it quaint to see boys walking around wearing straw boaters as part of their school uniform.

As we enjoyed the peace and tranquillity of the city centre, we had no idea that we were in an earthquake zone and could not have imagined the death and destruction wrought on the city by the earthquake of February 2011, or the slaughter of 50 innocent people in the attacks on two mosques on March 15, 2019.

We woke up on our second morning in Christchurch to torrential rain. We had not come prepared for rain but even umbrellas would have been of little use in a rainstorm like this. It eased off a little in the late evening and we nipped out to a nearby restaurant for dinner.

On day three we hired a car and drove to Akaroa about 50 miles south of Christchurch. We had planned to take a cruise to see the Hector's dolphins but the water was so rough that we dropped that idea. We walked around the magnificent harbour and noted that many of the seabirds were similar to our own back home – gulls, oystercatchers and terns. We decided to go for a walk in the woods. We had heard that early settlers had taken birds with them to New Zealand to try to prevent homesickness. We saw lots of chaffinches, greenfinches and all the usual birds we would expect to see on a stroll through the woods in Northern Ireland. We got an excellent view of a kingfisher. We heard a bird calling but didn't recognise the bell-like call. We eventually tracked down a small yellow/green bird with brown wings and brown tail. When we looked it up it was called, surprise, surprise, a Bellbird. When we got back into town the sun had come out so we decided to have lunch. Akaroa is renowned for its cuisine. There we were, on a bench overlooking the harbour, eating fish and chips!

Back in Christchurch we went for a walk in the Botanic Gardens but saw nothing except birds from home. We might have had more luck in the Willowbank Wildlife Park, in the north of the city, but the day lost to rain had knocked our schedule out.

One thing we were not going to miss out on was a trip to Kaikoura. Kaikoura is unique for its very deep waters that mix warm and cold ocean currents, forcing nutrients to the surface and attracting pods of up to 300 dolphins, sperm whales, fur seals and many marine species. A whale-watching excursion was to be the highlight of our trip to the South Island. Apart from these wonderful mammals we were promised that we would probably see lots of interesting birds like the black-browed mollymawk, Antarctic fulmar and the royal albatross.

We set off for Kaikoura with great anticipation early in the morning of day four. Consider our disappointment as we lined up to purchase our tickets when we spotted a large poster on the wall. The poster warned would-be travellers that, because of the stormy weather, the sea was rough and the likelihood of seasickness was pretty high. Neither of us are good sailors so the decision was easy. Mount Fyfee dominates the Kaikoura skyline. We would go bird watching there. We didn't have a lot of success with that either although we did see a grey warbler and a quail.

When we got back to Christchurch, we discovered that a group of the local judges had arranged to take us out for dinner. It was a pleasant way to end our visit to New Zealand. Next morning, we flew back to Sydney to catch our flight to Hawaii. We were in for another surprise – we had no hotel booked. We had arrived the day *before* we left New Zealand. We had forgotten about the international dateline!

We arrived at the airport around 11 pm and it was after midnight when we reached the hotel. They were not expecting us but said we could stay for the night and sort things out in the morning.

My contact in Hawaii had recently been appointed to the Supreme Court. His wife rang us next morning and invited us to meet them in their club that evening around 5:30 pm.

Our hotel was the Diamond Head Beach Hotel. We strolled along the sea front to Waikiki Beach. It was just as we imagined it – sparkling blue/green water, soft golden sands lined with palm trees. We had our photograph taken with parrots and then went diving for pearls. Well *diving* is not quite accurate! For a mere $10 you could pick an oyster out of a large tank using a pair of tongs. You then had to tap it three times with the tongs shouting "Aloha" and break the oyster open to find the most beautiful pearl ever discovered. Una decided to get three – one for herself, one for Sheila (our sister) and one for Lisa (my son's partner).

We went on to meet our friends in their club and were joined by two of their friends who had just returned from a holiday in Ireland and couldn't wait to tell us all about it. We spent a very pleasant evening together and arranged to meet next morning for a swim at Waikiki beach. The water was surprisingly cold but, once we were in, it was OK. Mike and Bonnie then took us to Lyon Arboretum where we had lunch before taking a guided tour of the park. We saw lots of birds but were not able to identify them. We searched some local bookshops for bird books with little success. We did however, find a book on Diamond Head – an extinct volcano, which we decided to climb next day.

In the park close to the hotel, we saw Red-Crested Cardinals, Java Sparrows, Red-vented and Red-whiskered Bulbuls, House Finches, Canaries, Japanese White-eyes, Golden Plovers and Zebra Doves.

It was very hot when we got to Diamond Head. The first half of the climb was easy. We then had 66 steps through a long dark tunnel followed by 99 steps through another tunnel before completing the climb via a spiral staircase. But the view from the top – looking out over Honolulu – made it all worthwhile.

We stopped for coffee on the way back into town and phoned Mike to enquire about Bonnie who had been unwell that morning. We arranged to meet for a swim. Mike offered to take me *outrigger canoeing* and offered to show me how to ride the waves. I hadn't been aware that, apart from being a Supreme Court Judge, he was also a fully qualified *outrigger* coach and spent a lot of his spare time coaching disadvantaged children. He then let Una have a go. Each of us was presented with *Outrigger T-Shirts* to show the world that we were now expert outrigger canoeists. Well, maybe "expert" is a slight exaggeration. Even a coach as skilled as Mike could not make us experts in one hour's tuition. Perhaps we could have benefited from just a *little* more coaching. We would have loved that. We had great fun and were sorry we weren't staying a bit longer. We got dressed, had dinner and said our goodbyes as we were leaving next morning.

On arrival at the airport, we discovered that security was extremely tight. First, we had to put our cases through a scanner which looked for foodstuffs. Then we were advised that our cases would have to go through a second and stronger scanner which would destroy any photographs or films we had taken. We were advised to take them out and place them in our hand luggage. This meant opening the cases and finding the items. We then had to accompany our luggage while it was scanned again. Eventually we saw our luggage off and went to check in. During all of the hassle with the suitcases Una had put the keys in her pocket instead of in her carry-on luggage and set off the alarm. She was taken to one side and searched again. Then her hand luggage was searched thoroughly. I got told off for advising on how they might speed things up a little. I was searched thoroughly too. They found my pill box and apologised because they "thought it was a lighter". They handed everything back and waved us through. We looked at one another but said nothing. They had failed to find *five* lighters which Una had bought in Hong Kong for her mother. So much for intrusive security checks!

It was late when we landed in Los Angeles where we had a 24-hour stop over. We caught a shuttle to our hotel at Long Beach and checked in. We were up early next morning for a walk along the shore. We saw lots of birds but weren't able to identify all of them because we didn't have a local bird book. We walked as far as *The Queen Mary* and took some photographs. Then we headed into town. We had lunch, looked around the shops and bought some souvenirs. We spotted some warm jackets which we liked and bought one each. I still wear mine working in the garden on cool spring or autumn days twenty years later.

Soon it was time to return to the hotel, pick up our luggage and head for the airport. Security was tight again but not as intrusive and the staff were more relaxed. Our flight to London was uneventful and soon we were queuing up again for the last leg of our journey. We had had an eventful six weeks what with the

World Congress, my election as President and our round-the-world trip. Nonetheless it was good to get home.

Brazil
(1986, 1997, 1998, 2010)

Copacabana here we come:

In August 1986, I set off for Rio de Janeiro to attend the World Congress of the IAYFJM. There was a large delegation from the UK organised by Dilly Gask and John Williams of the BJFCS so I was in good company. It was a huge congress with about 1500 delegates – the largest ever Congress organised by the IAYFJM. The majority of delegates were from South America with very large delegations from both Brazil and Argentina. I made a lot of useful contacts and became very friendly with a number of judges from the USA. I was invited to attend their 50th Annual Conference in Cincinnati the following year.

The weather was beautiful, as one would expect in Rio, and the lure of the beaches was very strong. John Williams and I, together with another English colleague and one from Scotland tried to get to Copacabana Beach every day during the two-hour lunch break. We were all of the same mind that following a huge breakfast and in anticipation of a huge dinner we could well skip a huge lunch. So, the four of us grabbed a taxi each lunch time and went for a swim. The beach was magnificent but the waves can be quite powerful. I recall once while swimming a wave caught me and turned me right over. I was being swept inward with my head scraping along the sand. I thought for a moment that I had reached the end of the road.

We had been warned to look out for pickpockets and told that we should not carry valuables with us – or leave our clothes unattended on the beach. So, we made a habit of leaving one person on guard while three went swimming. Once we left our colleague from the south of England on guard duty. It was a hot day and he had had a late night the night before. When we returned from our swim, we found him sound asleep – and some of the clothes gone. It was a sort of poetic justice when it turned out that the thief had stolen his clothes. He was a big man

– about 6 foot 3 inches – so perhaps the thief was a big man also, or maybe had a big client waiting for clothes. No one passed much remarks as we returned to the hotel in a taxi with one person just wearing his swimming shorts.

I made sure I didn't sleep when I was on guard duty. One day I lay on my back in the sun watching a plane circling overhead pulling a streamer advertising a concert by Lionel Ritchie. I had never heard of Lionel Ritchie but, in talking to my colleagues, discovered that he was promoting his latest album "*Dancing on the Ceiling*" (released in 1986). "*Ballerina Girl*", would later become my favourite song from that album.

When the various talks, presentations and workshops ended there was an opportunity to visit a facility for children – either a detention centre for young offenders or a children's home. I opted to visit a children's home. It was a sobering experience. All the children had been there for many years with little prospect of getting out. Usually when visitors like us arrived it was to select a child to take home. I wasn't able to find out if this ever led to a formal adoption or even how well the children were treated. It appeared to me that the centre didn't keep much by way of formal records. However, the children in the centre looked well cared for and there appeared to be a good relationship between staff and children. The children put on a show for us and made us feel very welcome. We were all given a present – something made by the children themselves. A girl in her mid-teens gave me an embroidered cushion and gave me a really tight hug as if I was her long-lost father who had turned up out of the blue. It saddened me to think of how vulnerable these children were.

The IAYFJM's World Congress always ends with a banquet. The Brazilians did us proud. There was a fabulous floor show, followed by a sumptuous dinner, followed by an opportunity to dance.

When the Congress was over many delegates, including myself, and all the British delegation, headed for Iguazu Falls. Iguazu Falls stands at the point where Brazil, Argentina and Paraguay meet. Stretching over a distance of almost two miles, it is the largest falls in the world. Numerous islands along the edge divide the falls into separate waterfalls and cataracts, varying from 197 to 269 feet in height. The number of these smaller waterfalls fluctuates from 150 to 300, depending on the water level. About half of the river's flow falls into a long and narrow chasm called the Devil's Throat. The border between Brazil and Argentina runs through the Devil's Throat –20% is on the Brazilian side and 80% on the Argentinian side.

There is a never-ending debate about whether the view from the Brazilian side is better than the view from the Argentinian side. I informed my British colleagues that I was taking a bus trip into Argentina to find out for myself. They informed me that it wasn't possible. When I asked why they said: "We are still technically at war. We couldn't get a visa to enter Argentina". (The Falklands war was in 1982 and only lasted a few months but diplomatic relations between the United Kingdom and Argentina were not restored until 1989). I had an Irish passport and didn't need a visa. I found the view spectacular from the Argentinian side but still could not decide which view was the best. The views are different, and equally beautiful.

I got an unexpected bonus from the trip. I have always been interested in wildlife. As we approached the falls, I spotted a giant monitor lizard, which I estimated to be about five feet in length, walking along the side of the road. Then, as I stood at the falls, I spotted a bird flying across from the Brazilian side. I trained my binoculars on it. I can still feel the surge of excitement as I realised that it was a toucan, a bird I had never seen in the wild before. I was amazed to see swifts darting in through the cascading water to reach their nests in the rocks behind. It seemed the most unlikely place to build a nest. But then, I suppose, it is totally secure from predators.

I stopped off in Paraguay on my way back to Brazil to get a few souvenirs.

Una had come with me to attend the IAYFJM's World Congress in Buenos Aires in 1998. She had taken ill when we stopped off in Rio and I had promised that, when my travel schedule allowed, we would go back to Brazil so that she could see the best of Rio and visit Iguasu. She had to wait twelve years before I got a chance to keep my promise. But all things come to those who wait.

On Friday September 3, 2010 we were heading back to Rio. By way of compensation for the long wait I decided we would go business class and use up some of my air miles.

We were flying via London and Lisbon. The London-Lisbon flight was an early morning departure so we spent the first night in London. Back to the airport again for our flight to Lisbon. After a short stop-over there we were in the air again and on our way to Rio. We had a 9-hour 30-minute flight ahead of us, but the seats were very comfortable, there was plenty of room to move about and the service was great. There was nothing of interest on TV so, after dinner, Una put her seat into the bed position and lay listening to music. I had some reading to catch up with. We landed in Rio at 9:50 and it took over 15 minutes to taxi to the gate. We had a 45-minute taxi ride to the hotel. The hotel wasn't quite what we expected. When we booked, they claimed to be four-star but three-star might have been over-generous. It is on a busy road so there was a lot of traffic noise. Still, it was only a few minutes' walk from Copacabana so we weren't complaining about that.

We were up next morning at 8:50 and down for a buffet breakfast. We then went for a walk and walked the full length of Copacabana beach. It was about 25^0C and overcast which suited us. Everywhere we looked people were walking dogs and most of them seemed to be poodles – standard, miniature and toy – just like our Lucy. Una said she wished we had taken Lucy with us. On the way back we decided to call in at the hotels to see if we could book rooms for when we returned from Iguasu but most were out of our price range and, in any event, all were fully booked. We had a wonderful lunch at one of the beachside cafes. It started to rain when we were there and they just let the sides down to stop the rain from getting in. We saw quite a few birds on the way back to the hotel: hummingbirds, vultures, doves, sparrows, kiskadees and frigate birds. We walked along Ipanema beach and saw Snowy and Cattle Egrets. We went out for dinner at 7 pm and stopped at another lovely beachside cafe.

We were up at 8 am next morning and had breakfast before going out for a walk. We again saw the standard and miniature poodles that we spotted the night before. Our taxi arrived at 10:30 so we checked out and headed back to the airport. It started to rain on the way but we didn't mind as we were leaving. We had a two-hour wait at the airport. Una wrote up her travel-diary notes and I worked on my notes for Liberia (I was due to fly there two days after we got home from Rio). We had a two-hour flight to Iguasu. Una watched out the window but didn't get a sight of the Falls. We landed at 3:45 and there was a car waiting for us. We had a 20-minute drive to the hotel through the national park. It had changed very little since we were there in 1998. Now that it is a national park you can only enter on a special tour bus or if you have a reservation in one of the two hotels situated within the park. The hotel looked the same. The Red-Rumped Caciques were there to greet us with their noisy calls, but the Coaties were nowhere to be seen. We checked in and were given a beautiful room with a huge shower and twin sinks in the bathroom. Our window looked out over the treetops – great for us as we were hoping to see plenty of birds. We could hear the water from the Falls just beyond the trees. We decided to go for a walk. It took a while to find reception because the hotel is so large. We booked a table for dinner and trips for Tuesday and Wednesday. Then it was down to the Falls for our first real view. They were as beautiful as ever but Una thought there was less water than when we were there in 1998. The Coaties were waiting to greet us and walked us down the path. One of them kept climbing into the bins to see if there was any food to be had. He was very funny and Una captured him on video. It was a wonderful walk and she took loads of photos. Then it was back to the hotel to change for dinner.

We were up bright and early next morning as we had booked a forest walk. We set off in a jeep but didn't get far before it broke down. We transferred to one of the tour buses and were dropped off at the starting point for the forest walk. Our guide, Marcus, was waiting for us. Marcus is a keen birdwatcher and so full of knowledge. We saw around 20 different species of birds and lots of beautiful butterflies. We arrived at a lookout point over a lake and saw a Caiman, Egrets, a White-Necked Heron and kingfishers. Then we took a boat trip along the Iguasu River before transferring to kayaks and paddling along the last part of the trip. Marcus offered to help us as my hands were too sore to manage it alone. We got soaked as the bottom of the kayak was full of water but it was great fun. Una bought a book in the shop and then we caught a bus back to the hotel. We left our clothes out to dry so they can get soaked again tomorrow. We had coffee and a sandwich for lunch and then headed back down to the Falls. The path was very busy but we managed to get past most of the tours. The Falls looked amazing and we took lots more photos. We bought some things in the shop and got a photo of the two of us with the Falls in the background. Then it was time to head back to the hotel. There were some small lizards along the path on the way back and then as we were nearing the hotel a black and yellow bird flew into the trees nearby. We stopped to see what it was and just then a Toco Toucan flew past us being chased by a smaller bird. Una was in her element. We had dinner at 7:30 as usual. We spent the evening writing up our notes before going to bed.

Our tour next morning wasn't until 11 am so we decided to go for a quick walk after breakfast before getting ready. We took the path to the right which wound through a forest. We could hear noises in the trees and at first thought it was the Coaties but it turned out to be a small troop of monkeys. We also managed to see some lizards and a Squirrel Cuckoo. We headed off at 11 am and this time the jeep didn't break down. We had a short drive through the forest and then a walk down to the docks to get on the boat. We left our cameras in a locker as we were told they would get very wet. This meant we didn't get any photos. As it turned out, we could have taken the cameras so long as we had kept them covered when close in to Falls. The boat trip was really good. We went out over the rapids and then the skipper turned the boat into one of the falls and we all got soaked. It was like a torrential shower and we loved it. We soon dried off again in the heat so after the boat trip we decided to go out to the visitors' centre at the entrance to the park. While we were waiting for the bus a beautiful butterfly took a liking to me and Una got some good photos. It was called an 88 as it had 88 marked in its wings. The centre wasn't up to much but we got a nice coffee and bought some presents for people back home. Back at the hotel we decided to have a dip in the pool. It was freezing but fun. We wrote some post cards and had coffee with the Red-Rumped Caciques noisily flying around overhead. A large moth decided to join us for dinner that evening and sat on our table while we ate. Una got some more photos.

We decided to take it easy the following day and hadn't booked any trips. After breakfast, we walked down to the Falls. There was a strong breeze which caused a lot of spray and as a result there were many beautiful rainbows. The Dusky Swifts had arrived back and it was amazing watching them hanging on tight to the rocks as the water washed over them. On the way back along the path we spotted the point where we had our 'shower' the previous day. We waited for a boat to appear so we could get some video. We were going to have a coffee but decided to leave the gifts we had bought in our room. We were so glad we did. On the way back to the pool we could hear something calling and it wasn't the Red Rumped Caciques. We went looking for it and discovered it was a Toucan but not a Toco. It was very cooperative and allowed us to get some photos. We spent the rest of the afternoon by the pool drinking coffee, writing cards and going for a dip. The water was surprisingly cold but it was all right once we got in. We found it hard to believe that this was our last night here. On our way to dinner, we asked if we could have a later check out next morning but, as all the rooms hadn't been allocated by then they said they would let us know later. When we returned to our room after dinner to get packed, we found a note waiting for us to say we could check out at 1 pm.

At breakfast next morning we could hear birds calling around the grounds so when we were finished eating, we went looking for them. We found some beautiful blue and green finches and a hummingbird. We watched them for a while and then went to get the camera. They had gone when we got back but we waited for a while and then the hummingbird returned followed by the finches. We got lots more photos. We also saw some Mockingbirds, Rufus-Collared Sparrows and a House Wren. We dragged ourselves away to finish packing and then had coffee by the pool. The large butterflies must have been feeling the heat as some of them had come down to the pool to drink – more photos. On the way to reception to check out we saw an interesting tree that we hadn't noticed before so we stopped to have a look. We spotted the most beautiful finch we have ever seen – blue, green, black and orange. A Mockingbird family was also setting up home in the tree.

We checked out and then had lunch by the pool. Our taxi arrived at 2:30 and we were on our way. Una said that, in her view, Iguasu is the most beautiful place on earth. She hoped that, one day, she would get an opportunity to come back.

Our flight took off on time and we were back in Rio at 5:40 pm. We had a 45-minute drive from the airport in one of the airport taxis. The taxi driver pulled over on a street corner and told us we had arrived at our hotel. There was no sign of our hotel! The driver didn't speak much English. I got the number of our hotel from my notes and told the driver to phone them. It turned out our hotel was at the other end of the street so he took us there. We checked in and this time we had a better room. We had a lounge, bathroom and bedroom so we had somewhere to watch TV or read. It was beside the room we had originally. We unpacked and then went down and booked a tour for the next day. We were going to visit the 'Christ the Redeemer' and the Sugar Loaf. We had a walk on Copacabana beach and stopped for an orange juice. We wandered through the market on the way back to the hotel and bought some gifts for folks back home.

We had difficulty opening our room door and had to go down to reception for assistance. We had the same problem next morning on returning from breakfast. The staff seemed to think the problem was with us! We asked at reception for the international dialling code for home so we could phone about Lucy but we couldn't get the phone to work.

Our bus arrived at 9:15 and we were off for our city tour. Our first stop was at the European Cathedral which didn't impress us. We had an opportunity to walk around the area where the annual Carnival is held before going on to the football stadium where the 2014 World Cup was due to be held. Next stop Christ the Redeemer. It was crowded as usual but we did get some great photos including one of us shaking hands with the 'Christ'. A good time to break for lunch! After lunch we headed to the Sugar Loaf. All the Cable Cars were full and it was difficult to get any good photos. However, the views from the top were amazing and the coffee we had was excellent. We were back in the hotel at 6:30 and this time we had no problem with the door. Time to chill out and then it was early to bed.

We had left the next morning free to allow time for relaxation. After breakfast we went for a walk on Copacabana. Una was wearing her ¾-length trousers and decided to walk in the water. I joined her. Some of the waves were bigger than we expected and we got a bit wet but it was good fun. We walked the full length of the beach and then had to head back as Alyrio, a colleague who sat with me on the Executive Committee of the IAYFJM, was to pick us up at 1 pm. He was looking well – considering he was approaching his 90th birthday! We took a taxi back to his house as he had given up driving. Alyrio's wife, Thais, had prepared a beautiful lunch and their daughter, daughter-in-law and granddaughter joined us. We had a very enjoyable afternoon. We left around 5 pm and their daughter showed us to a shopping mall close by the apartment. It was a lovely mall but there were no souvenir shops. We found a Starbucks and had a coffee before getting a taxi back to the hotel. We had a rest and then went for a walk along the beach. We stopped at one of the beach side cafes and Una ordered a pizza. A young man came along and asked if he could show us his work. We said no but he wasn't going to give up. He sat down and got out his paints and began to paint on a tile. It was actually very good and Una decided to buy it. We were not sure if we would be able to get it home as the paint might take a long time to dry. (In the event we did and Una still has it.) After dinner we wandered about the market for a while but didn't see anything worth buying.

Next morning, after breakfast, we rang the kennels back home to enquire about Lucy. We were assured that she was fine. We booked a couple of trips for the following day and then asked for advice on buying souvenirs. We were referred to a shop not far from the hotel. We headed off to the gift shop and got the last few things we needed. We stopped at one of the beach cafes for a fruit juice before going back to the room for a rest. At 3:30 we decided to go for a swim. We walked along the water's edge and a few times the waves came up and splashed us but all our gear was in plastic bags so it didn't matter. We walked to the end of the beach. On the way back I decided it was time for a dip. We had just put our shoes down to get organised when a big wave came and nearly washed them away. We moved our clothes up the beach a bit and then I dived into the waves. Una took some photos of me getting tossed about. Then it was her turn. I was so busy taking photos that I didn't notice another big wave coming. Our clothes were almost washed away. A couple of people stopped to help me gather things up while Una raced out of the water. Between the four of us we managed to get everything – soaking wet, of course. Luckily, I was holding the camera when the wave came. So it had come to no harm. We made our way slowly back along the beach as the wet shoes were hurting my feet. We got back to the hotel none the worse and had a shower. After dinner we wandered through the market as Una wanted a few last-minute gifts. Then it was back to the hotel. We had now mastered the art of getting our eccentric door to open so we didn't need anyone to come and help us.

Down for breakfast at 7:30 the next morning and on the road at 8:40 for a jeep tour of the forest. There was a Chinese lady (from Houston) in the jeep and she was very funny. She jumped when an insect flew past her and our guide Lynda reminded her that she was going on a tour of the forest. It was a really interesting trip, even though we would have preferred walking rather than driving. We stopped at a few places for photos. Lynda was very knowledgeable about everything we saw and good fun. We stopped at a gift shop and bought a few things. We were heading on to the area where we were promised we would have a walk when our driver spotted something in the trees. It was another Toucan and a new one for us. It flew off before we could get any photos but at least we saw it. A short walk through the forest with Lynda pointing out different plants to us. The lady from Houston was jumping at every insect that flew past and every noise in the trees. At the end of the walk a family of coatis came to meet us. We were advised not to feed them but the lady from Houston decided she would give them some biscuits. Of course, as soon as they heard her opening the packet, they came running towards her. She almost had a heart attack. The tour was over all too soon and we were dropped off at our hotels. The temperature had risen to 33^0C – the hottest day since we arrived. We went to one of the beach cafes for lunch and then back to the hotel for a rest as we were going to a show that night. It was too hot and noisy to rest so we decided to do our packing since we were leaving the following day. Our bus arrived at 7:40 pm and we were taken to a restaurant which specialised in meat and BBQ. Not the best place for a vegetarian! However, there was a wide selection of food in the buffet and I was able to get as much as I wanted. We were sitting with two ladies from South Africa and they were telling us that they were also going to Machu Picchu and

Patagonia. We were able to tell them all about our trips. The show at Plataforma was excellent. It was very colourful and the dancers were amazing but it wasn't a true Brazilian show. It had a very international flavour. It lasted about one and a half hours and we were back at our hotel about midnight. We were in no hurry up in the morning as we didn't need to check out until 3 pm.

We couldn't leave our hotel without a final piece of drama. We went back to our room after breakfast to find our door wouldn't open. Una spotted one of the cleaners and asked her to let us in with her passkey. Her key wouldn't work either. I called reception and Walter came up and, guess what, he couldn't open the door. He called maintenance. They sent someone up to change the battery. That didn't work. At least they knew now that the problem wasn't with us. The door lock really was faulty. The maintenance man eventually got the lock working again.

We went out for a final walk on Copacabana and had lunch at one of the beach cafes. We returned to our hotel, had a shower, packed the last of our stuff and checked out. Walter urged us to come back soon and, with a smile, promised that he would make sure we got the same room. On the way to the airport Una noticed that the traffic had come to a standstill on the way into Rio (luckily, we were going out) and the taxi man switched on the news. Apparently, there had been an accident with two buses and it looked like no one was going anywhere for a while. Una told the taxi man that he should stay at the airport until the traffic had cleared and he said he would do just that. We had no problems checking in and we were told that there wasn't a problem checking the cases right through to Belfast. However, as we picked up our tickets, I noticed that our luggage tags suggested that we were only going to Portugal. I went back to reception and was assured that the luggage tags would be replaced. Who knows where our luggage will end up! We went through security and into the business lounge. We eventually took off 45 minutes late. The flight was uneventful but a bit bumpy. We landed at 9:15 am and were in the air again at 11:15 am. We arrived in London at 1:25 pm and caught a bus to Heathrow. We left London at 6 pm on the final leg of our journey home. We landed in Belfast at 7:05 pm. It was hard to imagine it was all over now and we would be able to get Lucy the following morning. Of more immediate concern was our cases hadn't turned up. We were told that it isn't possible to book luggage right through if you are changing airports in London. We could only hope that our luggage had made it to London and would be delivered to our home the following day.

We were up early next morning and out to the kennels for 8 am forgetting that they didn't open until 8:30 am. Lucy was happy to see us and we got a good telling off all the way home. She just sat on Una's knee and howled at us, as she always does when she comes from the kennels. However, all is forgiven as soon as we get home. We couldn't do much that day as we were waiting for our luggage. We did some shopping and then just stayed about the house waiting for a phone call. The cases eventually arrived about 6 pm and we were able to get unpacked and sort out the washing. Thankfully, all would be in order for my mission to Liberia.

Canada
(1981, 1982, 1987, 2004)

My son Liam was an only child. One might expect that our world would have revolved around him. But this was not the case. Our world revolved around his mum who had been an invalid all his life. What we did, where we went always depended on how his mum was. I was concerned that Liam might feel neglected because his wishes never got priority. In 1980 Liam was approaching the end of his primary school career and was due to sit the 11+[12] exam to determine what secondary school he would go to. I sat down with him to discuss which secondary school he would like to go to and to explain how the exam result might increase or limit his choice. I said that if he did well in the exam and got his first choice of school, I would take him somewhere nice. He could choose where. He asked what would happen if he didn't do well. I said we would go somewhere nice anyway to celebrate the end of his primary school career. He did well and got his first choice of school.

Liam's favourite TV programmes were American and he was fascinated by New York after reading a story about it in one of his comics[13]. He said he would like to go to New York. Initially the plan was that his mum would come too but

---

[12] The 11+ is a selective entrance examination for secondary school, used by both state-funded grammar schools and many private schools in England and N Ireland to identify the most academically-able children. The exam is taken towards the end of Year 5 or beginning of Year 6 of primary school. The children will be more than 11 years old but less than 12 – hence the name "11+".

[13] Red Dagger No.9 – "Terror In The Tall Tower" – a reprint of an old Hotspur story, in which the main character becomes the unexpected owner of a Manhattan skyscraper and has a number of run-ins with organised crime figures. Published in the early spring of 1981.

the doctor ruled that she was not fit to travel. We agreed to postpone our trip until the following summer.

With summer '81 fast approaching and Liam's mum still unfit to travel it was decided that Liam and I would go on our own. Leaving home on June 24 we would fly to North America, via London, and visit New York, Toronto, San Francisco and Los Angeles. I tagged on an extra day in London to the end of our trip, but in the event, we couldn't go either sightseeing or comic-hunting there because of the Brixton riots. We were due back home on July 12. Because, in this book, reports on countries visited are dealt with in alphabetical order, I will report on our visit to Toronto here (under Canada) and report on New York, San Francisco and Los Angeles later (under USA)

After a few days in New York (see p237) we arrived in Toronto late in the morning of Sunday, June 28, 1981. We would be on the road again at 6:30 am on the Tuesday to catch our flight to San Francisco, so we had a day and a half in Toronto. On the first day we went up the CN Tower, at that time the world's tallest freestanding structure (only a radio mast in Warsaw, which needed propping up, was taller). It was a glorious sunny day, so we had superb views. Being Sunday, a lot of shops were closed so we didn't do much shopping. We did, however, happen upon a large underground shopping complex, where you can escape the summer heat or winter cold. That evening we went for a swim in the hotel pool. Then my nephew Teddy came and took us to his home in Pickering, Ontario, where his wife, Beatrice, had prepared dinner for us.

On Monday morning we headed for Niagara Falls. We decided to take a trip on the "Maid of the Mist". Liam was annoyed when he was made to wear a black raincoat but, as we approached the Falls, he realised that he would have been soaked to the skin without it. We took lots of photographs. We both agreed that the Canadian Horseshoe Falls were more impressive than the American Falls.

But, my outstanding memory of my first visit to Canada is that I found it too hot for comfort. I expected the temperature to be in the mid to high 20ºs C, but, unusually, it was in the mid 30ºs C.

I was looking forward to sunny California with temperatures in the low 20ºs C. But more of that later. This chapter is focusing on Canada.

We were back in Canada the following year, this time for an extended holiday. We spent 10 days in Montreal where we stopped with my niece Ann, her French-Canadian husband, Lou, and family, followed by four days in Toronto with Teddy and Beatrice.

Lou was keen to show off the booming cultural metropolis which Montreal had become under the guidance of visionary Lord Mayor Jean Drapeau. Drapeau had been Lord Mayor for a total of 29 years spanning the Sixties and Seventies, when many of the city's biggest projects happened, such as the 1967 World Fair (best known as Expo 67) and the 1976 Summer Olympics. 62 countries hosted pavilions and it is estimated that 50 million people visited Expo 67[14]. Many installations had disappeared over the years but much remained. A trip to Île Sainte-Hélène and Île Notre-Dame was top of the "must see" list Lou had prepared for us. His eight-year-old daughter, Angela, adopted me as her favourite uncle and volunteered to be my special guide.

As we wandered around La Ronde[15] we came to a stall where the stall-owner guaranteed to guess my age. He claimed to have a 99% success rate. If he failed, we could claim a free teddy bear. He failed miserably and we went on our way with Angela hugging her new friend. She knew two other stalls where they would offer to "guess your age". She went home with a clutch of teddy bears.

It was very hot in Montreal, about 35°C. I burn very easily and wouldn't normally dream of going swimming in an open-air swimming pool. But Angela was insistent that I go with her. Eventually I gave in. She introduced me to all her friends. All of them were used to the sun and were as brown as berries. I stood out like a sore thumb. Suddenly I realised why Angela had wanted me to come. She was so proud to have an uncle who was unique. None of her friends had an uncle with such pale skin.

Liam said that it was too hot for him to go swimming. He opted to stay "at home" with his cousin Gavin and read his comics. And yet, when I got back from the swimming pool, he decided that it wasn't too hot for us to go tramping around Montreal searching for two comic shops he had been told about. He was in high spirits when we located both of them. We returned to Ann's to relax and enjoy a long cold drink.

We were intrigued with Montreal's "Underground City" (officially called "RÉSO"). The name applied to a series of interconnected office towers, hotels, shopping centres, residential and commercial complexes, convention halls,

---

[14] 1967 was the centenary of the Dominion of Canada, a key stage in it becoming independent.

[15] La Ronde is an amusement park built as the entertainment complex for Expo 67, the 1967 world fair. The local baseball team is called the Montreal Expos.

universities and performing arts venues that form the heart of Montreal's central business district, commonly known as "Downtown Montreal".

The network is particularly useful during Montreal's long winters, when well over half a million people are estimated to use it every day. I must admit that, for me, it was an excellent place to escape the scorching summer sun. The network is largely climate-controlled and well-lit, and is arranged in a U-shape with two principal north–south axes connected by an east–west axis. Combined, there are 32 kilometres (20 miles) of tunnels under 12 square kilometres (4.6 square miles) of the most densely populated part of Montreal. In total, there are more than 120 exterior access points to the network, not including the sixty or so Metro station entrances located outside the official limits of the RÉSO, some of which have their own smaller tunnel networks. The network is completely integrated with the city's entirely underground rapid transit system, the Montreal Metro. Liam was fascinated by the Metro system and, before we left, could name every station, in proper order, on the Green Line, from the city centre to Honoré-Beaugrand station, where we got off for Lou and Ann's, and the short yellow line which took us from Berry de Montigny to Île-Sainte-Hélène, where La Ronde was located. Gavin kept testing Liam on the stations, but never caught him out! The metro was amazingly cheap for children, just 25 cents to go anywhere in the city! Liam and Gavin made full use of it.

Another "must see" was Notre-Dame Basilica. The interior of the church is amongst the most dramatic in the world and regarded as a masterpiece of Gothic Revival architecture. The vaults are coloured deep blue and decorated with golden stars, and the rest of the sanctuary is decorated in blues, azures, reds, purples, silver, and gold. It is filled with hundreds of intricate wooden carvings and several religious statues.

We took a day trip to Québec City where we saw the Québec provincial parliament and the area near the palatial Château Frontenac hotel, where many celebrities and historical figures have stayed in the past. Unlike largely bilingual Montreal, Québec City was essentially all French-speaking. That summer Liam got his first opportunity to practice French outside of the classroom.

Further afield we visited the Basilica of Sainte-Anne-de-Beaupré. This magnificent basilica overlooks the Saint Lawrence River, 30 kilometres (19 miles) east of Quebec City and is one of the five national shrines of Canada. It is an important Catholic sanctuary, visited by about a half-million pilgrims each year.

Another day trip was to Ottawa. We went to see the Canadian Parliament building where Liam and Gavin had their photograph taken with the Mounties and were allowed to pet the horses. We didn't stay around too long as it was even hotter than in Montreal.

It was hotter still on a day trip to the countryside when Lou brought us to visit his sister. Luckily, Liam and Gavin had the use of two paddle boats which kept them busy, and cool, for a while.

Our holiday in Montreal was over all too soon. We thought of flying to Toronto but Teddy and Beatrice insisted on driving to Montreal to pick us up. Gavin came with us to Toronto, or rather, Pickering, Ontario, where Teddy and Beatrice lived.

Top of Liam's "things to do" was to go on another comic-hunting trip, this time in downtown Toronto. I can still remember the startled look on the faces of passengers on the bus as Liam pointed out the window screaming "Queen's Comics". He also wanted to visit the Science Centre and Planetarium. Finally, Teddy took us on a one-day visit to the local nuclear power plant. After four pleasant days in Toronto/Pickering we flew to Boston for the last three days of our holiday before heading home.

Our next visit to Canada was in August 1987. I was in Cincinnati, Ohio, attending the 50th Anniversary Conference of the (US) National Council of Juvenile and Family Court Judges (NCJFCJ). My wife, Bernie had died in July 1986. I didn't like leaving Liam alone so soon after losing his mum. I asked him if he would like to come with me. After Cincinnati, we visited courts and juvenile facilities in New York, Washington, Baltimore, Houston, Los Angeles, Las Vegas and Minneapolis. My niece, Ann, invited us to spend a few days with them in Montreal, if we had time, after this hectic tour of the US. We were happy to accept.

Lou and Ann had moved house since we were here last, but Honoré-Beaugrand was still the nearest metro station. On this visit the temperature was a more pleasant 27°C! We visited La Ronde again, did some more sightseeing and a bit of shopping. The three days passed very quickly and we were homeward bound.

The Canadian Association was one of the most active national members of the IAYFJM. They hosted the World Congress in Montreal, 17-22 July, 1978 and a number of other international conferences over the years. Most of the conferences were held in Montreal but I had visits to Quebec City, Ottawa, Toronto and Vancouver. I will comment briefly on one other visit to Montreal.

In 2004, I was working with the UN Human Rights Centre in Geneva in the preparation of a Training Manual on Human Rights for Judges and Lawyers. I had joined forces with Professor Jean Trépanier from the University of Montreal to write a chapter on the administration of Juvenile Justice. We had been corresponding for some time by email as we worked through the chapter but, as we were coming towards a conclusion, we decided we needed to get together to put the finishing touches to it. I agreed to join Jean in Montreal. The timing wasn't the best. It was late January and January in Montreal can be severe. I was expecting it to be cold so I had brought warm clothes with me. I wasn't expecting it to be − 27°C and, when the wind chill factor was taken into account, the real temperature was probably closer to − 37°C. I have never been so cold in my life. I survived by wearing my pyjamas under my clothes and stuck it out for two weeks until we completed our task.

Chile
(2003)

I flew to Santiago, Chile, in 2003, to present a paper in Santiago University before going on to a seminar in Tongoy. I had a great view of the city as we came in to land, nestling in the centre of the *Santiago Basin*, a large bowl-shaped valley consisting of broad and fertile lands surrounded by mountains. A blue haze hanging over the city reminded me that Santiago is prone to smog.

There was a lot to see in Santiago but very little time to see it. There was no time to explore the pedestrian malls or try out one of the many cafes or restaurants. My host took me on a whirlwind walking tour of the centre – pointing out the statue of the Virgin Mary at San Cristobal Hill, one of the main symbols of the city, the Metropolitan Cathedral, one of the few remaining examples of colonial architecture, the church and convent of San Francisco. The tour finished at The Pontifical Catholic University of Chile (UC). The UC is one of Chile's oldest universities and one of the most recognised educational institutions in Latin America. The Faculty of Education, where I gave my lecture, ranks 33rd worldwide – according to the QS Ranking. In 2018, the university was rated 1st University in Latin America.

I had always wanted to visit Valparaiso – probably because it is mentioned in so many seafaring songs. So, when I found out that I had a free day between the university presentation and the seminar, I decided to spend it there. One of the judges volunteered to take me and show me around. Then I got word that the Chilean President was signing the *Optional Protocol on the Rights of the Child* that day. Could I attend in my official capacity as President of the International Association of Youth and Family Judges and Magistrates? It was a difficult decision. I wanted so much to see Valparaiso. I might never be back in Chile. There would be lots of dignitaries attending the signing of the Optional Protocol.

My attendance, or nonattendance, would barely be noticed. My devotion to duty got the upper hand. I went to the signing.

My next stop was Tongoy, a small town on the Pacific coast, about 250 miles north of Santiago, 26 miles south of La Serena – Chile's second oldest city. The town is situated on a rocky promontory between two beautiful white sand beaches – Socos, which is 4 km long and Grande, 26 km long. The town has two main residential areas: the *Peninsula*, comprised mainly of summer homes and the *Pueblo Bajo*, where the majority of permanent inhabitants live. The Conference organisers had rented the summer homes for attendees. I had a beautiful chalet to myself. I could happily have spent a couple of weeks there but, as usual, time was pressing.

My friends in Chile were concerned that I was subjecting myself to too much stress because of the hectic pace of my life. One of the judges had a holiday home in southern Chile. I was invited to spend a week there to relax. I was so disappointed for both my hosts and for myself because I had to turn the offer down. My calendar was full and the schedule tight. I had seminars looming in Scotland, Switzerland, Mexico, Poland and South Africa. I had to get back home.

China

(1992, 2005)

In 1992 the Supreme People's Court of The People's Republic of China (PRC) invited the IAYFJM to co-sponsor an international conference in Shanghai. I was asked to present a paper on Juvenile Justice in Northern Ireland.

As usual, when travelling to a country I had never been to before, I wanted to garner some background information before setting off. The following is a summary of material provided by Professor Wei Long[16], a Supreme Court judge in China, and a member of the Executive Committee of the IAYFJM.

Juvenile Courts in China

Prior to 1984, there were no juvenile courts in China. In December that year The People's Court of Changnin District in Shanghai established the first judicial tribunal to deal exclusively with juvenile cases. By the spring of 1988, courts of justice in 21 districts and counties of Shanghai had established juvenile courts, and The Intermediate People's Court of Shanghai had established a juvenile court of second instance, making it possible for the first instance and the final instance of juvenile cases to be heard within the jurisdiction of the juvenile court.

Developments were not confined to Shanghai. With the encouragement and guidance of the Supreme People's Court, the experience and methods of the juvenile tribunal spread quickly. By 1988 there were more than 100 tribunals set up throughout the country. By June 1992 this number had increased to 2763, with a staff of 11,008, so that all juvenile cases in China could be heard in juvenile tribunals.

The Supreme People's Court set down the following guidelines: Juvenile tribunals shall be composed of at least one full-time professional judge and two

---

[16] This was the situation as of 1992

assessors. The assessors will be part-time, elected or specially invited, and will be unpaid. At least one of the assessors will be a female. Judges dealing with juvenile delinquents must be very good at judicial business and enthusiastic about working with young people, having some expertise in psychology, criminology or pedagogy. Because of the demands of the work, courts will set up classes to train the judges and give them the relevant knowledge. The assessors should have a good understanding of young people and have some expertise in education such as teachers, or people who have been engaged in youth work for a long time. The assessors will have equal rights with the judges in trying cases.

I was surprised at how closely this reflected our system in Northern Ireland! The Supreme People's Court guidelines continue:

The judge should interrogate the juvenile defendant informally, face to face, before the trial so that the defendant has the chance to present his/her opinions on the alleged offence fully and freely. In court, handcuffs will not be used and judicial police will not be present; the defendant will be allowed to sit, and will be allowed to respond to the charge and present his/her opinions freely. The court will be less formal than the general criminal courts, with the judges, the juvenile defendant and other litigation participants sitting opposite one another and on the same level. The judge will ask questions and give explanations in a mild tone, using simple language and will try to create a relaxed atmosphere, an atmosphere of equality so as to narrow the psychological distance between the defendant and the judge. The judge will not tolerate any rebuke, sarcasm, or threat toward the juvenile defendant. News broadcasting from the court will not be allowed – film, TV or radio programmes. It is forbidden to publish the juvenile defendant's name, address, photograph and any material that could be used to make the juvenile publicly identifiable. On a finding of guilt, the juvenile defendant and the legal representative will have the right to appeal to a higher juvenile court. If the finding of the juvenile court is upheld there is no further appeal but presenting a petition is allowed.

Development of a Family Court[17]

The success of juvenile courts highlighted an obvious anomaly in that the protection of the substantive rights and procedural rights of the delinquent

---

[17] This is still part of the 1992 briefing.

juvenile can be heard in the juvenile court but not the protection of the fundamental rights of the non-delinquent juvenile. Litigation concerning the rights and interests of juveniles in need of protection (except inheritance and maintenance in the civil law) is heard in courts of general jurisdiction. Cases where parents are guilty of an offence on the grounds of abuse or abandonment of their children are heard by the general criminal court. There is a need to set up a multi-functional and fully-protecting juvenile judiciary to deal with all affairs concerning the juveniles, including abuse and abandonment by parents. Furthermore, the jurisdictional scope of the "narrow range" philosophy of juvenile tribunals obviously misses the opportunity for the juvenile court to assist those juveniles whose behaviour is giving cause for concern but who are not delinquent having committed no offence. There is also a need to expand the "narrow range" philosophy and allow the use of the forces of comprehensive treatment in the form of the centripetal nets of education and correction centred in society for the protection of the delinquent juvenile to be extended to the unhealthy juvenile. Such a court would be on the lines of a family court. Two district courts in Shanghai, Tianning District Court of Changzhou City, Jiangsu Province, and some district courts in other provinces have tentatively established comprehensive juvenile courts to accept not only cases of juvenile delinquency, but civil, economic and administrative cases concerning legal rights of juveniles so as to meet the requirement of giving juveniles all-round judicial protection. Shanghai led the way in the development of juvenile courts and now looks as if it is leading the way in the evolution of family courts in China.

Now that Juvenile Courts had been established in all districts of China and thoughts were turning towards the establishment of Family Courts, the Supreme People's Court of the PRC deemed it a good time to bring together experts from East and West to share experiences. Young people account for about one third of the world's population. They are the future of mankind. Therefore, the issue of prevention, adjudication and rehabilitation of juvenile delinquency is a topic of common concern and interest. The Supreme People's Court decided to host an international seminar to be held in Shanghai from November 14 to November 16, 1992 and invited the IAYFJM to co-sponsor it. The IAYFJM members supported the proposal and agreed that Executive Committee members should be in attendance. I decided to have a few days in Hong Kong before attending the seminar.

Two Resident Magistrates (RMs) from Northern Ireland had recently taken up posts in Hong Kong. I understood their move followed threats on their lives by the IRA. I knew one of the RMs well as his parents lived next door to me. I decided to stop off on the way to Shanghai and visit him in his court. This would give me an insight into Hong Kong's justice system. At the same time, I could visit an old school friend who was working for the University of Ulster, and was based in Hong Kong with a view to encouraging Chinese students, and possibly students from other Asian countries, to come to Northern Ireland. This was a fast growing and very important market as Chinese students would pay three times the fees which local students would pay. The University of Ulster was ahead of the field in having their man in Hong Kong.

I stopped with my friend for a couple of days and he showed me the highlights of the city. I found my visit to the court very informative and enjoyed a chat over dinner afterwards with the two RMs. The next day I headed for the airport to catch my flight for Shanghai only to find that it had been delayed. As I wandered around the airport I bumped into Lucien, a judge from Canada who served with me on the IAYFJM executive. We managed to rearrange our seats so that when we eventually took off, a couple of hours late, we were able to sit together. The time passed quickly as we chatted away and we were touching down in Shanghai "in no time".

Shanghai

We were queuing up at passport control when we were approached by two police officers. One said: "You come with us". Perhaps rather foolishly, considering the reputation of the Chinese police, I said: "No. We have to get through Passport Control and then collect our luggage". The police officer repeated his command and again I said no. Then he said: "You come with us, *now*". The change of tone made me decide that discretion was the better part of valour. We jumped the queue for passport control and were led off towards a police car. As we were being bundled into the car I asked: "What about our luggage?" The police officer replied: "We will look after your luggage. Now we must get you to the banquet". We were horrified. We knew that a banquet had been arranged where we and our executive colleagues would be hosted by The Supreme People's Court of Shanghai. We were dressed in jeans and T-shirts. How could we go to a banquet? We insisted that we *must* go to the hotel first so that we could change. The police officer again said "No. The President (of The

74

Supreme Court) is waiting. He insists that you come now". In normal circumstances, had we arrived at the banquet dressed as we were it would have been regarded as a personal insult to the President. We would have risked arrest. But here we were, dressed in jeans and T-shirts while everyone else was "dressed to kill". Back home we would have expected a warm round of applause, but not here – just smiles from our colleagues. We were ushered to our seats. As soon as we were seated everyone was asked to stand up again and the President and senior judges came in and took their seats at the top table. As Lucien and I enjoyed a wonderful meal we tried to forget our embarrassment. We could laugh about it afterwards and it was a story to tell.

I had never been in Shanghai before, or indeed in China. So, everything was new and exciting. One image that sticks in my mind is that of men constructing a motorway. There were hundreds of men armed with picks, shovels and wheelbarrows – not a bulldozer or dumper-truck in sight. It was very obvious on my second visit, 13 years later, that such squads of labourers had been stood down in the race to modernisation and replaced by machinery. Where were they? Had they been absorbed elsewhere in the workforce or had they joined the ranks of the unemployed?

Another image that sticks in my memory is the number of bicycles. Katie Melua sang "There are 9 million bicycles in Beijing". I didn't see as many bicycles in Beijing as I saw in Shanghai. I can't say how many there were but can confirm that there were millions of them. I can still see in my mind's eye a dual carriageway running through the centre of the city. It was about four traffic lanes wide on either side but not a car or lorry in sight. The carriageway was packed, on both sides, with cyclists – so tightly packed that I wondered how it was possible to cycle without bumping into someone and ending up in a tangle of bicycles.

On another occasion, I was anticipating a tangle of cars, with much more deadly consequences. We were on our way to meet the President of Shanghai's Supreme Court – the man who could not be kept waiting. Once again, we were in a police car but the traffic on the motorway was practically at a standstill. There was no way the police car could get through – even with the siren blazing. One police officer jumped out and opened the barrier to allow us through to the other side of the motorway. I wondered if we were going back to find an alternative route. But no. The driver set off at full speed up the wrong side of the motorway against the flow of traffic, siren blazing and lights flashing. The police officer in the passenger seat was leaning out the window waving to oncoming cars to get out of our way. My heart was in my mouth as I wondered how far we would get before we met a car head on because the driver hadn't realised in time that we were coming the wrong way. I don't know how we survived but we did. The important thing from the police point of view was that we hadn't kept the President waiting.

The seminar was attended by judges, attorneys, public prosecutors, professors and officials from 30 countries and regions including Argentina, Belgium, Canada, France, Germany, Indonesia, Iran, Italy, Kuwait, Mauritius, Mexico, Mongolia, the Netherlands, Nigeria, Northern Ireland, Pakistan, Romania, Russia, Saudi Arabia, Singapore, Spain, Sri Lanka, Thailand, Tunisia, United Arab Emirates, United Kingdom, United States, Venezuela and Hong Kong. It was an outstanding success.

Mr Bogdanesen, President of the Supreme Court of Romania, spoke for us all when he said, "Although the seminar cannot solve all the problems of juvenile delinquency in all countries, the sharing of experiences, and the exchange of views, gives us confidence and courage knowing that we are all working together toward a common goal. Individually we all make contributions to the prevention, adjudication and rehabilitation of juvenile delinquency. Working together we can ensure that juveniles everywhere will have a better future."

Opportunities to meet socially are just as important as the formal meetings. I have already mentioned the banquet hosted by The Supreme People's Court of Shanghai on November 13 (when Lucien and I arrived late). There was a cruise on the Huangpu River on the night of November 14. The following evening there was a reception hosted by the Deputy Mayor of Shanghai, followed by a visit to see "the most famous acrobatic troupe in the world – Shanghai Acrobatics". Next morning participants had the opportunity to visit a rehabilitation school and the Shanghai High People's Court.

I was impressed with the rehabilitation school we went to. The atmosphere was very relaxed with an easy relationship between staff and residents. This reinforced my belief that children in the justice system in China are treated well. Still, I was surprised that the residents were given the opportunity to learn English and they all wanted to try their English out on us.

I had already noticed that young people in Shanghai were keen to practice their English. I was up early each morning for a walk in the park. I was not surprised to find lots of people out exercising. I was surprised to be approached by young people who rightly guessed I was an English speaker. They told me their name and asked me to tell them mine; they asked me where I was from and if I was here on holiday. Then they thanked me very much and bade me good day. I was very impressed.

The seminar ended on November 16, but our Executive Committee (of the IAYFJM) had been given permission to stay on for two extra days to hold a meeting. Much of the discussion centred on the preparatory work for the next World Congress which was due to be held in Bremen, Germany, from August 29 to September 2, 1994.

Beijing

A Government representative informed us that we were invited to Beijing for a holiday, the Chinese Government would cover the cost. It was an offer we couldn't refuse. However, we had a Supreme Court Judge from Russia on the Executive Committee. She said that she would have to contact the Russian Embassy for permission to go with us. She rang the embassy. The Ambassador said "No" and ordered her to return to Russia immediately. This tall, austere Supreme Court Judge, who could speak very little English, burst into tears. I never expected to kiss a Russian judge but she was so distressed that I gave her a big hug and kissed her on the cheek. I said we were sorry she couldn't come and looked forward to seeing her again at our next meeting.

On the morning of November 19, we were flown to Beijing. Beijing was very different from Shanghai. The first surprise was when someone spotted a McDonalds. I was not a fan of McDonalds but, for most of the group the Big M sign was like a long-lost friend waiting to welcome them. They were making a beeline for it.

Apart from McDonalds, Beijing at that time had a very third-world feel about it. The pavements around Tiananmen Square were lined with artisans offering all kinds of services, there on the pavement. It was possible to have your hair cut, get measured for a suit (or a dress for the ladies), go to the dentist, have your shoes polished or a new pair made. It appeared to me that you could buy almost anything you could possibly want.

The most exciting part for me was being in Tiananmen Square itself and recalling the events of June 3/4, 1989. The Tiananmen Square protests were student-led and received broad support from city residents. The students occupied the Square for seven weeks. Hardline leaders ordered the military to enforce martial law. Troops with assault rifles and tanks inflicted thousands of casualties on unarmed civilians trying to block the military's advance. The Chinese government condemned the protests as a "counter-revolutionary riot", and all forms of discussion or remembrance of the events within China were prohibited. So, it was not a topic we were likely to be able to raise with our Chinese hosts. But I had watched it all on television and it was a strange feeling walking around the square and recalling all that had happened.

The Chinese imperial palace known as the *Forbidden City* stands at the top of Tiananmen Square. It served as the home of emperors and their households, as well as the ceremonial and political centre of Chinese government for almost 500 years. It was declared a World Heritage Site in 1987, and is listed by UNESCO as the largest collection of preserved ancient wooden structures in the world. I was not going to miss the opportunity to see at least some of its 980 buildings, marvel at the traditional Chinese palatial architecture and view the extensive collection of artwork and artefacts dating back to the Ming and Qing dynasties. I had never, in my wildest dreams, pictured myself here.

The Great Hall of the People dominates the Square. It has been described as "one of the truly grand modern structures of Beijing. Its green and yellow glazed-tile roof, magnificent portico and colonnades, and rows of pines and cypresses create a look both solemn and immense." It is even bigger than the Forbidden City. It is the political hub of Beijing and home of the National People's Congress. Nowadays it is open to the public but in 1992 it was not. It was a great honour therefore when Mr Ren Jianxin, President of the Supreme People's Court of PRC, invited us to a banquet there. In his welcome address, President Ren said he greatly appreciated the contribution our Association was making to protecting and promoting the healthy growth of minors.

The service did not commence until the President and senior judges had taken their seats at the top table. Once we got started the atmosphere, at least at our table, was very relaxed – perhaps a little too relaxed. It was an excellent meal. I can't recall how many courses. I just recall that desert was served and the green tea poured and we were still chatting away. Suddenly the Master of Ceremonies clapped his hands and we all had to stand up. The President and senior judges stood up and filed out. We were about to sit down again to eat our desert and drink our tea when we were all ushered out. The party was over. We had learned a valuable lesson – eat first and talk afterwards!

Over the next few days there were trips to the Great Wall, the Forbidden City, The Summer Palace and the Temple of Heaven and the Beijing Opera. Our hosts were determined that we be given every opportunity to sample some of the finest Chinese cuisine.

That was the end of our official visit to China but a few of us had decided we would like to visit Xian (about an hour's flight from Beijing) to see the *Terracotta Army*. This is a collection of terracotta sculptures depicting the armies of Qin Shi Huang Di, the first Emperor of China. It is a form of funerary art buried with the emperor in 210-209 BC and whose purpose was to protect the emperor in his afterlife. It has been estimated that there are over 8,000 soldiers, 130 chariots with 520 horses and 150 cavalry horses – the largest pottery figurine group ever found in China. It is an amazing sight – a fitting way to finish off a marvellous trip.

I had two more things to do before leaving Beijing.

I asked my hosts for permission to visit the teacher-training department of the university to have discussions regarding the possibility of setting up links with St Mary's Teacher Training College back in Belfast (where I lectured). They provided me with a police car and driver.

My discussions at the University proved fruitful and the professor in charge was keen to form such links. I was to get a very different reaction from the principal in St Mary's, when I got back home. He looked at me in horror when I said I had negotiated a potential link with Beijing University. How could I even imagine that a Catholic institution like ours would want to have links with a Communist institution? I was ahead of my time. Today all institutions vie for the opportunity to establish such links!

My second task was much less controversial. At that time, I had a cousin whose hobby was collecting flags from different countries. She was excited when she heard I was going to China – hoping that I would bring her back a flag.

When I had finished my discussions in the University my driver asked me if there was anything else, I would like to do or see. I asked him if he could bring me to a shop where I could buy a small Chinese flag. He told me that he was not aware of any shop in Beijing which sold small flags. There were few tourists coming to Beijing in those days so there was no demand for such items. But he had a solution. There was a flag on the front of the bonnet of the police car. He removed it and handed it to me. Not what I expected from the stereotypical image of a Chinese police officer as garnered from the media!

I was concerned that he might get into trouble when he returned to the police barracks minus the flag. He assured me that he would not get into trouble. The Commissioner would be pleased to learn that I respected China's culture and traditions so much that I wanted one of their flags.

I was back in China in 2005. Beijing had changed a lot in the thirteen years since my first visit. All the stalls along the streets were gone. No more leisurely strolls around Tiananmen Square. The roar of traffic was incessant. China was no longer a third world country. It was now a developing nation and the development was amazingly fast.

Shanghai, too, had changed and was now a very modern city with skyscrapers. All too often, rapid growth leads to a hideous mix of height and shape and colour with individual buildings standing out like a series of sore thumbs. The planners in Shanghai had achieved a harmonious, spectacular skyline.

The Chinese have a great respect for "position". At the time of this visit, I was President of the International Association. I was met at the airport by two supreme court judges. We bypassed the queue for passport control. I was taken directly to the VIP lounge to wait while my passport was cleared and my luggage collected. Then I was escorted to an official car with diplomatic plates, one of the judges carrying my bags, and rushed off to my hotel.

One of the judges who met me was Xui Li, the new Chinese representative on our Executive Committee, my old friend Wei Long having retired. Xui Li said she had arranged for me to meet with Wei Long the next day.

Wei Long and I had a long chat over several cups of green tea. He was so different now to when we first met in Sion, Switzerland, some years previously, and he told me that Tiananmen Square was what would now be called "fake news". I don't know if it was because he was now retired or whether he had learned to trust me over the years. It was probably a bit of both. He told me what Tiananmen Square was really like. We talked about the Cultural Revolution launched by Mao Zedong (1966 through 1976). Senior officials and the professional classes were accused of taking a "capitalist road". Millions of people were persecuted in the violent factional struggles that ensued across the country, and suffered a wide range of abuses including public humiliation, arbitrary imprisonment, torture, sustained harassment, and seizure of property. Wei Long told me how the *Red Guards* had entered the courts and publicly humiliated the judges, many of whom were dragged off and made to do manual work in the countryside. He wasn't sure why he had been allowed to retain his position as a judge – perhaps because he had never spoken out publicly against Chairman Mao. It was refreshing to know that Wei Long now trusted me enough to be so open about things.

Xui Li asked me if I would like to go for a walk on the Great Wall of China. Who wouldn't? When we were alone on the wall, she said she really wanted to seek my advice. She wanted to be sure that no one would overhear our conversation. The Government was concerned about international criticism of China's use of the death penalty. There are more people put to death for a range of offences in China than in most of the other nations in the world taken together. Xui Li, now a Supreme Court Judge, had been asked to chair a group of Supreme Court judges whose task was to come up with a solution for reducing the number of annual executions. She told me that they had identified one reason for the steep rise. Some years previously, in an effort to cut the burgeoning crime rate, the High Courts had been made the final court of appeal against the imposition of the death penalty. She was minded to recommend that the Supreme Court be made the final court of appeal as it had been previously. She estimated that this would reduce the number of executions by at least 20%. What did I think? I said that I was opposed to the death penalty, as she already knew. I realised that abolishing the death penalty in China was a non-starter, at the moment. I said I thought that most reasonable people would agree that a cut of 20% was a very positive commitment and that I imagined that it would be welcomed by Human Rights Organisations and by the UN. I commended her on her initiative. A next step might be to reduce the number of non-violent crimes which carried the death penalty. I felt that her step was lighter as we completed our walk. I had never expected to be strolling along the Great Wall of China discussing such an important human rights issue.

I had already had discussions with my colleagues in the IAYFJM about China and the death penalty. They were totally opposed to the idea of a reduction in the number of executions. They wanted total abolition. I unexpectedly bumped into Lord Wolf (recently retired British Lord Chief Justice) in Beijing. Without breaking any confidences, I asked him for his views on encouraging China to reduce the number of executions pending total abolition. He fully supported my view that reduction was a very important first step.

The precise number of executions carried out in China is regarded as a state secret so the exact number is unknown. According to Amnesty International the official tally was 2,148 in 2005. This dropped to 1,591 in 2006, a 25% decrease. I like to think that maybe my intervention played a role in bringing this reduction about.

In most countries which still retain the death penalty, capital punishment is reserved for the most serious of crimes such as aggravated murder. China had a long list of 68 capital crimes the majority of which are non-violent. In 2011, the National People's Congress Standing Committee adopted an amendment to reduce the number of capital crimes from 68 to 55. Later the same year, the Supreme People's Court ordered lower courts to suspend death sentences for two years and to "ensure that it only applies to a very small minority of criminals committing extremely serious crimes". Was my walk, and talk, on The Great Wall bearing fruit?

Cuba

Havana (2005)

In 2005, I received an email from *The Law Development Research Institute (IDID) and the General Attorney's Office of Cuba,* announcing that *The V International Meeting on Legal Protection of the Family and the Minor* would take place at the Havana International Conference Center, Cuba from November 15-18, 2005. I was invited, as President of the International Association of Youth and Family Judges and Magistrate, to address the conference. The title of my paper would be: *Implementing the Convention on the Rights of the Child – The Challenge for Judges.*

I had never been to Cuba so my first task was to find out how best to get there. US sanctions meant that there were no direct flights from any US airport. I was surprised to find out that I could fly London – Toronto – Havana with Air Canada. Once I had my "Cuban Tourist Card" I was all set to go.

I like to do my homework on any country I am going to for the first time. I knew that Fidel Castro had ousted General Fulgencio Batista, Cuba's American-backed president, in 1959. For the next two years, officials at the US State Department and the Central Intelligence Agency (CIA) attempted, unsuccessfully, to push Castro from power. In April 1961, the CIA launched what its leaders believed would be the definitive strike: a full-scale invasion of Cuba by 1,400 American-trained Cubans. The invasion took place at The Bay of Pigs. It did not go well. The invaders surrendered after less than 24 hours of fighting. The "invasion" is sometimes referred to as "The Bay of Pigs Fiasco".

I knew about the Cuban Missile Crisis of 1962 and how President John F Kennedy had faced down Soviet Premier Nikita Khrushchev. In the aftermath of the crisis, the US announced an embargo blocking companies from trading with the island. Cuba became isolated from the world. The embargo, undoubtedly, had a major impact on the Cuban economy. So, how were the Cubans coping?

I learned that all Cuban residents had free access to health care in hospitals, local polyclinics, and neighbourhood family doctors who served on average 170 families each – one of the highest doctor-to-patient ratios in the world. However, the health system had suffered from shortages of supplies, equipment and medications caused by the ending of Soviet Union subsidies in the early 1990's and the US embargo. Despite these difficulties, infant mortality rate was lower than many developed nations and the lowest in the developing world.

Cuba had an excellent education system. The national government assumed all responsibility for education, and there were adequate primary, secondary, and vocational training schools throughout Cuba. Education was compulsory at primary and secondary level. It was free at all levels – including university.

I was looking forward to my visit. I wouldn't have much time for sightseeing but my hotel – the Meliá Habana – was in the centre of the city, as was the International Conference Center. Havana, population about two million, has been described as three cities in one: Old Havana, Vedado and the newer suburban districts. Old Havana, with its narrow streets and overhanging balconies, is the city centre. With the limited time available to me, I intended to focus on Old Havana.

Old Havana takes you back in time with its vibrant architectural mix. The Castillo de la Real Fuerza, a fort and maritime museum, is a fine example of 16th-century Spanish colonial architecture. The National Capitol Building is an iconic 1920's landmark. The imposing Catholic Catedral de San Cristóbal features a baroque, ornamental facade and two bell towers. It is regarded as Havana's finest example of 18th-century Cuban baroque. The cathedral stands within the area of Old Havana that UNESCO designated a World Heritage Site in 1982. The positions of the original Havana city walls are the modern boundaries of Old Havana.

The cathedral contains a number of sculptures, paintings and frescoes. I just mention a few I particularly liked. There is a sculpture of Saint Christopher, Patron Saint of Havana, which dates from 1632. There are copies of paintings by Rubens and Murillo on the altars in the side chapels. There are three fading frescoes – *The Delivery of the Keys*, *The Last Supper* and *The Ascension* – by Italian artist Giuseppe Perovani above the main altar. There is also a canvas by Perovani of the Virgin of the Immaculate Conception, Patroness of the Cathedral.

84

The Plaza de la Catedral is one of Old Havana's most beautiful spots. The square, together with Plaza Vieja and Plaza de San Francisco, were within walking distance from my hotel and from one another. This is a great place to walk around and buy souvenirs. At least, it would be if you could do it in peace. Every time I went out for a walk one or more young women would join me offering to be my guide. When they refused to accept that I didn't need a guide I would check my watch, say I had a meeting to go to and head back in the direction of the hotel. If I went in somewhere for coffee, they would be waiting for me when I came out. Once I went into a restaurant for lunch, placed my order and took out some papers to read. A young woman joined me and asked if I would like to buy her lunch. I said no. I had some work to do and needed peace and quiet. I have attended conferences all around the world and had never experienced harassment like this before. Not that the young women were nasty, they were all most polite. They just would not accept that no meant no. I guess the US embargo was impacting on their business, as on so many others.

Another unintended consequence of the US embargo was that, over the years, Cuba became a rolling museum of American classic cars. With no new American cars or parts, the Cuban population had to make do with what parts and vehicles they already had, mainly 1940's and 50's era models. These vehicles are kept running through the ingenious use of hand-built, improvised parts. Many cars have been almost completely rebuilt over the years. They may look original but are far from it. Amazingly there are still plenty of vehicles with original engines and well looked after interiors and exteriors. These cars are reserved for the exclusive use of tourists. Old Havana is beautiful, with its brightly coloured buildings and milling streets. But it is the classic cars which give the city its unique character. Let us hope that the easing of the embargo by President Obama will not be the death knell of the classic cars. Their loss would inevitably change the city's character. President Trump may inadvertently come to the rescue – not that he wants to rescue anything in Cuba but it could well be an unintended consequence of his actions.

Ecuador
(2014)

From discussions with Una about my trips abroad I was aware that she dreamed that, one day, she would visit the Galapagos Islands. So, when the opportunity arose to visit Ecuador, she jumped at the chance of coming along. Her dream came true when we flew to London on the first leg of our flight to Quito (via Madrid). We stayed overnight in London and were back at the airport at 4:15 am for our flight to Madrid. There was a bit of a queue for check-in but it moved fast and we had no trouble. Our bags were checked right through so we didn't need to worry about them until we got to Quito. The flight to Madrid was uneventful and we landed at 10 am. Our next flight took off at 1 pm and we had a 10-hour 30-minute flight to get to Quito.

Quito and the Cloud Forest

The landing in Quito was one of the smoothest I have ever known. You wouldn't have known the plane had touched the ground. We landed at 4:30 pm Quito time and it took a long time for us to clear customs. We didn't mind as we thought the luggage would be waiting for us when we got through but it wasn't. It was s a new airport (only opened about a year) and they still seemed to be having some problems. It took us about two hours from landing until we were finally ready to leave the airport. At least Marta, our guide was waiting for us. We now had a very long drive (40 km) into the city as the new bypass out to the airport was still under construction, and it was rush hour. We were glad to get to the hotel. We checked in and Marta said she would pick us up at 9 am. We had a quick bite to eat in the restaurant and then it was time to get to bed. It was now 21 hours since we had had any sleep and the altitude was beginning to get to us. Quito is 2,800 metres above sea level and neither of us does well at altitude.

I didn't sleep much. I was very sore (a mixture of Parkinson's and arthritis) and the altitude made it difficult to breathe. Una slept a bit better. We got up at 7 am and had a light breakfast – toast, cheese and fruit. Then we went for a short walk. We found a square near the hotel that had some trees which were full of doves and sparrows. Marta arrived at 9 am as promised and we set off on our ½-day city tour. The tour started at the large statue of the 'winged' Virgin Mary which stands overlooking the city. The views were amazing. Quito is a very long thin city – 50 km long and only 5-10 km wide and sits in "the avenue of the volcanoes". We spent some time there taking photos and enjoying the view. Then it was back to the old town to continue our tour on foot. Chocolate is one of Ecuador's biggest exports. We visited a chocolate shop where we bought some 100% and 80% chocolate. Quito is a very religious city so the tour included a visit to some of the many churches including Conjunto Monumental San Francisco. The church dates back to the 1570's and was devoted to St. Francis, since the Franciscan order was the first to settle in the area. After our visit here we stopped for a drink in Plaza San Francisco. Our tour took in another church before finally finishing in the park we walked to that morning. It turned out that this park is known as Plaza de la Independencia. It is the central square of the city and is the location of the offices of the country's Executive. Its main feature is the monument to the independence heroes of August 10, 1809, a date remembered as the First Cry of Independence.

After a short break back at the hotel we headed up to the square again and had some lunch. Marta had recommended a couple of restaurants and we tried one of them. Una had chicken and I had trout. We both enjoyed our choice. There was also a gift shop here so we got some post cards and a T-shirt. The local representative from Auldey Travel (our tour company) had arranged to meet us at the hotel at 4:30 pm to have a chat and see if everything was ok. We spent the rest of the afternoon relaxing in our room. We managed to get connected to the Internet so we sent messages home to let everyone know we were safe. Of course, Una had to check up on Kiki! Apparently, she was having a ball. The rep arrived at 4:40 pm and stayed until 5:20 pm. She had papers for us and went over everything as our times had changed since we received the original itinerary. We would have an early start in the morning as our guide would be at the hotel at 5:30 am to take us to the Cloud Forest. We went back up to the cafe in Plaza San Francisco to have some cake and coffee but they only had chocolate cake so we had a drink of juice instead.

We were up next morning at 4 am. We were told there would be a light breakfast available but when we went down there was nothing left out. Luckily, we had ordered a packed breakfast to take with us so we ate some of that and had coffee. Our guide, José, arrived at 5:30 am and we had an hour and a half drive to Yanacocha reserve where we were going for a birding walk. The drive out of the city was hair-raising with a tour bus trying to put us off the road just because he wanted to get past. Once we left the city, we left the main road and took a very bumpy track. We arrived at Yanacocha at 7 am and had a wonderful three-and three-quarter hour walk there. José was a wonderful guide and I don't know how he managed to spot so many birds in the trees. One of the things he asked us about was whether we had ever seen a Condor. We had tried on different occasions to see one but were never successful. We were back at the bus at 10:45 am and were just heading off for the Lodge when José told the driver to stop the bus and said, 'there's a Condor'. Sure, enough there was one floating high up on the thermals. It was an amazing end to a wonderful walk. We now had a very long and bumpy drive down the mountain to get to the lodge. We stopped for a packed lunch at 12:15 pm and arrived at the lodge at 2 pm.

Tandayapa Bird Lodge is set right in the Cloud Forest. They have a patio on which they have placed many hummingbird feeders. There were hummingbirds everywhere and they didn't mind us at all. They kept buzzing round our ears – one came to check if there was any nectar in my ears. We spent a while watching them before going in for coffee and a rest. José was waiting for us when we went back up to the lounge. It turned out that because this was the off season, we were the only people staying in the hotel. We spent some time talking about the hummingbirds on the balcony and our itinerary for tomorrow. Another early start as we were to leave at 5:30 am and had an hour's drive before going on a birdwatching tour. Dinner was served at 6:30 pm and, after that, we decided to have an early night.

We were up next morning at 5 am and left the lodge at 5:30 am. We reached our first stop at 7:00 am. We had been given a packed breakfast and ate that before spending some time in a beautiful garden watching for birds. We saw lots of birds, including Toucans. We also saw what we thought were hummingbird hawk moths. We couldn't get any photos of them as they wouldn't stay still. After our walk in the gardens, we had a walk through part of the forest. The walk was hard going with steep and slippery paths but we managed. We saw some interesting birds here including a Plumbeous Hawk which Una spotted sitting in the trees after José had walked past it. Maybe he was just being nice and leaving something for us to find! After this walk we drove to a different part of the forest and had another walk. The path here was muddier but wasn't quite as steep so we enjoyed it more. Then it was time to eat our packed lunch. The place we had stopped was beside a house and a little dog kept coming over to see if there was anything for him. He looked a little like a poodle who needed a good combing. We felt sorry for him. Una gave him all the sausage but José and the driver kept chasing him. Now it was time to head back to the lodge. I asked José about the possibility of getting stamps for post cards so on the way back we called in to Mindo, a tourist town not far from where we were staying. The heavens opened just as we arrived and both José and I got soaked for nothing as no one seemed to know what we were looking for. We headed back towards the lodge but instead of turning right to go to the lodge we turned left to follow the route we had come along the previous day. By now I was tired and didn't want to do any more walking but as it turned out we didn't have to. José was scanning the forest as we drove along, looking for a 'Cock-of-the-Rock' – one of Ecuador's most iconic birds. I don't know how he saw it as it was on the other side of the ravine but suddenly, he told the driver to stop and back up. He jumped out and got out his scope. I watched where he trained the scope and just managed to see a bright red speck before it flew off. Lucky for us it flew over to our side of the ravine and we managed to get quite a good look at it through the trees. We were so happy as this was one of the birds at the top of our list but we never imagined we would actually see one. We finally got back to the lodge at 5 pm very tired but very happy. Just time to clean our muddy boots and have a much-needed cup of coffee before dinner. We suggested to José that we would have a break in the morning and not go out early because he had a really bad cold and was struggling but he insisted we must go as he didn't want us to miss anything. So, no lie-in for us in the morning. Back to our room to shower and go to bed.

We were up next morning at 5 am and down to meet José at 5:45 am. We had a short walk into the forest in the dark to a hide. They leave food here to attract moths and then as it begins to get light the birds come down to eat the moths. I felt really sorry for the poor moths! We stayed for about an hour and five different birds turned up – Streak Capped Tree Hunter, Strong Billed Woodcreeper, Immaculate Antbird, White Throated Quail Dove and Chestnut Capped Brush Finch. Then it was back to the lodge for breakfast. It was nice to have a proper breakfast as we hadn't had one since our first morning in Quito.

After breakfast we spent some time photographing and taking video of the hummingbirds. They are amazing little birds and move so fast. Then it was time for our last trip into the forest. We drove to the ridge overlooking the lodge and walked along this road for some distance with the van following along behind. It was a pleasant walk and we saw many interesting birds including the Plate-Billed Mountain Toucan and an Andean Pigmy Owl. José was amazing at spotting the birds when we could see nothing. At 11 am, I said we should head back to the lodge as the cloud was settling in around us and I didn't want to have wet clothes to carry back to Quito. We had a bit of time to relax before lunch and bought a bird guidebook. It would really help in identifying all the birds we had seen. We set off from the lodge at 1:30 pm and had a three-hour hair-raising drive back to Quito. This time we didn't take the long bumpy route over the mountain as we had on the way to the lodge but followed the main road around the mountain with its many twists and turns and all too many crazy drivers who seemed to enjoy overtaking on blind bends. At one stage we came upon an overturned lorry. Luckily, it had overturned on the road and hadn't fallen down into the ravine. Hopefully everyone was ok. It was causing a bit of a tail back and some drivers behind us couldn't wait and came speeding up just managing to squeeze between the lorry and the other traffic and didn't seem to care that someone could be hurt or dead. It took us an hour and a half to get into Quito and then we spent the next hour and a half getting across town in the rush-hour traffic. We got to our hotel, which was out beside the airport, at 4:30 pm. At first, they claimed we weren't booked in. However, they did eventually find our names so all was well. We said goodbye to José and our driver and went to our room. It was a lovely room with two huge double beds and a massive bathroom.

We arranged to have dinner at 6 pm and had just decided to take a walk around the grounds when there was a thunderstorm. We arranged for the taxi to take us to the airport at 8 am as we thought 9 am was a bit late even though we didn't have far to go. We went back to the room to relax and have a shower. Then we had a knock on the door and the manager said we had a phone call. It was the local representative checking to see how we had got on in the forest and telling us there was no point in heading to the airport before 8:25 am in the morning as the boat rep would not be there before that.

We were wakened early the next morning (5 am) by dogs barking, an alarm going off and cockerels crowing. We got up at 6:30 am and decided to have a walk in the grounds of the hotel as the sun was shining. There wasn't much to see but it was a nice walk. After breakfast it was time to finish packing and have another walk in the grounds before getting the taxi to the airport. The trip only took 20 minutes but the driver couldn't speak any English. We didn't know which airline we were flying with as the rep from the boat company was supposed to be meeting us. A lady came over and asked us our names but we weren't with her group. We got out our papers but it didn't help much and so I went over to talk to the lady again as she spoke English and might be able to help. She was standing with a few other people and one of them happened to be our rep. Happy days. The taxi driver wanted to charge us $15 but we told him it had already been paid. He wouldn't let us go until he phoned the hotel to check. We got through baggage control and check in with no problems and then just had to wait until the gate was announced.

The Galapagos Islands

We took off at 10:30 am for a 35-minute flight to Guayaquil. We remained on board and had a 40-minute wait for other passengers to join us. We took off again at 12 noon for a 1 hour 35-minute flight to the islands. We landed at 1:30 pm and spent some time getting through customs and passport control. We picked up our luggage and went to find the reps from the boat. We met an English couple, now living in Spain, who were also on our boat. Our guide was waiting for us and took the luggage to be loaded into trucks and taken off to the boat. We had to wait for a bus to come back and pick us up. Then it was off to the port to get the dingy out to our boat. At the port we got our first look at some of the local wildlife as the seals were all around the dock. There were twelve of us in total boarding the boat plus four people who had stayed on board after completing a five-day cruise. This would be their second five days. We had iced tea on arrival and then lots of briefing before lunch. Then it was back to our cabins to get unpacked and settled in.

All was going fine until I started feeling unwell. Una had brought motion sickness tablets with her but when reading the instructions on the plane on the way to the islands we discovered I couldn't take them because of my Parkinson's. By now we had set sail and we were stuck on the boat for the next five nights. There was an excursion to Santiago Island that afternoon and we went on that as I was glad to get off the boat for a while. We had a 'wet' landing so we had to put on our water shoes and carry our hiking shoes with us. Una set her bag and camera down on a rock to change her shoes and a few of the local sea lions decided they were going to help themselves. Tanya the guide had to chase them. We had an interesting excursion and saw Sallylightfoot Crabs, Penguins, Lava Lizards, Marine Iguanas, Pelicans and Frigate Birds. After about an hour we had to get the panga (dingy) back to the boat. This time we got a bit wetter as everyone was climbing into the panga with very wet feet and the seats were soaking. We decided we should put on swim gear the next time we had a wet landing. We were back on board at 6 pm and had a meeting with the captain and crew at 7 pm. We all had to introduce ourselves and have a drink. Then it was time for dinner. I decided to skip dinner so Una had to go on her own. We had told Tanya about the seasickness and she got me some ginger tea to drink as this can help settle the stomach. She had also tried to contact a doctor regarding the tablets but hadn't managed to get hold of anyone and now we were out of mobile range. Una had a small amount of dinner and then came back to the room. She decided to take the motion sickness tablets as she thought it was best that she was feeling OK to look after me if needed. I was very tired so we went to bed at 9 pm. We had an overnight sailing to get to our next stop. We prayed it wouldn't be too rough and that I would get some sleep.

Day two on the boat – only four more to go. At least I slept all night and wasn't feeling so bad. We went down for breakfast. I just had some toast. The others were all asking how I was and offering me different tablets which they had brought with them. Then a lady from Bermuda (Alison) spoke to Una. She said that her husband can be a bad traveller and she always carries motion sickness wrist bands with her for him. She asked if I would try one. Una said that she thought I would try anything at this stage. Alison went and got one and put it on for me. She advised me to continue to drink the ginger tea as well. We set off for a panga ride around Vicente Roca Point, Isabela Island with no walking involved. We saw Penguins, Blue-Footed Boobies, Nazca Boobies, Sealions, Fur Seals, Sea Turtles and Marine Iguanas. It was an interesting trip and we enjoyed it. Then it was back to the boat to get changed to go snorkelling. I didn't go as the water would be too cold and Una can't swim so we stayed on the boat. I had a rest and Una sat on the balcony watching the turtles floating by and writing up her notes. Lunch was at 12 noon. I had a little soup and a vegetarian meal which the chef prepared for me. We both decided to skip dessert. Now we had a two-hour sail to the next stop. We sat on the balcony and had a very smooth crossing to Espinosa Point on Fernandina Island. We landed on the island, at around 4 pm, for a walk.

The lava formation was amazing. It reminded me of moonscape. We saw lots of Marine Iguanas and one of their nesting sites as well as lots of sea lions. It was necessary to watch where you stepped as there were lots of Lava Lizards running around and at one stage, we saw a small snake wending its way through the Iguanas. We also saw Flightless Cormorants, American Oystercatchers, a Lava Heron and a Wandering Tattler. We were making our way along the path back to the panga but a large sealion was blocking our route. Tanya decided it was best not to disturb him so we took an alternative route. I had been asking if we might see a Galapagos Hawk on any of the trips and Tanya said it was possible. Suddenly she called me and pointed to a hawk sitting in the tree just above our path. The sealion had done us a favour as we may not have seen the hawk otherwise. Back to the boat and time for showers before dinner.

After dinner we had a briefing session with Tanya about the next day. We would be moored here all night and then have a short sail in the morning before breakfast. One of the group (Glenn – an English businessman, retired and living in Spain) had been taking video while out snorkelling and let us all see some of that. It was great to see all the underwater activity. We went back to our cabin about 9 pm feeling tired but more relaxed. I had enjoyed the day and hadn't been feeling sick at all. Alison was a life saver and we will be forever grateful.

Just as we finished breakfast next morning someone noticed that a pod of dolphins had come into the bay to join us. Everyone piled into the pangas and set off to watch them. The dolphins seemed quite happy to stay around so Tanya said that anyone who wanted could put on a wetsuit and go swimming. However, when people entered the water, the dolphins moved to the other side of the bay. Each time we followed them they would move to the opposite side. We had great fun watching them from the panga. Some of the others got good video shots of them. By now we were well behind schedule for our walk on Tagus Cove, Isabela Island, so we reluctantly left the dolphins in peace and set off.

We landed at 8:30 am and it was already very hot. We had quite a long hike up the mountain passing Darwin Lake with views of the volcano. The views were amazing but I think most people were struggling in the heat. We were glad to get to the top and admire the views while enjoying the cooling breeze. No one wanted to leave because we knew we had the long walk back to the bay. A barbeque, and a few cold beers for those who drank, would have gone down a treat! But no such luck. We saw Flightless Cormorants, Penguins, Brown Pelicans, Small Ground Finch, Smooth Billed Ani, Galapagos Mockingbird and

Galapagos Finch. We were glad to get back to the boat out of the heat. The group were now going snorkelling and Tanya asked if Una and I would like to go kayaking. How could we resist? It was great fun just paddling out in the bay watching the turtles gliding by.

Back to the boat and time for a shower before lunch. It turned out that there was a barbeque for lunch so the group were happy and most had a cold beer to wash it down. Then we had a two-hour sail to Urdina Bay, Isabela Island, for our afternoon walk. En route to the landing point our panga's engine stopped working so the other one had to turn round and tow us to the beach. When we landed, we spotted a Galapagos Hawk sitting on a sign at the top of the beach. We edged closer but he wasn't concerned about us and happily sat on while we all took photographs. Then Tanya pointed out the turtle nesting site just behind him and we realised why he wasn't moving off. He was waiting for dinner to emerge.

The walk was a nice gentle walk around part of the island and we saw Giant Tortoise and Marine Iguana. We were making our way back to the beach when a very large Giant Tortoise came wandering along the path. Tanya said he was around 100 years old and if we all stood still, he would just pass on by without being bothered by us. However, he decided to turn to the right before he came to us. It was just amazing to see something so big that has lived for such a long time. The Hawk was still sitting on his perch on the beach waiting for dinner but now he had been joined by a Lava Gull (the rarest gull in the world) also hoping for dinner. The poor hatchlings won't stand much chance with these two about.

Now most of the group were going swimming and Tanya said she had brought a wetsuit and snorkel for Una to have a go. Una enjoyed it as we were on the beach and the water wasn't deep. She said she couldn't see much as it was a bit rough and the sediment was being stirred up but she didn't mind. It was fun being in the water. We had to do two trips back to the boat as our panga still wasn't working but it meant that some of us got to stay on the beach for a little longer. The boat set sail as soon as we were all on board as we had to make it to our next stop before morning. There was just time for showers before dinner. Considering there was such a small kitchen on the boat the chef cooked up some amazing food. Una was sure we must be putting on weight.

We docked at our next stop shortly after dinner and then it was time for the evening briefing. Marcus (one of the crew) had got some amazing video of the dolphins that morning so we watched that before heading to bed. Another early start in the morning as breakfast would be served at 6:30 am.

It was worth getting up early the next morning just to see the amazing sunrise over Elizabeth Bay. After breakfast, we had a panga ride through the mangrove forest. Landing there is prohibited so we had to stay in the boats. However, it was a very interesting trip and we saw lots of wildlife – Pelicans, Blue Footed Boobies, Flightless Cormorants, Penguins, Small Ground Finch, Yellow Warbler and three types of herons: Lava, Striated and Great Blue. We also saw sea lions, turtles and Golden Cow Rays.

Back to the boat for drinks and snacks and then a two-hour sail to get to our next stop. We sat out on the balcony watching the world go by. When we docked the rest of the group went out snorkelling but we stayed on board. Una sat out on the balcony again writing up her notes and watching the group snorkelling. She said that suddenly something appeared to fall from the sky and there was a great splash as it entered the water. She watched to see what it was and a Blue Footed Booby surfaced with a fish in its beak. It didn't swallow the fish but seemed content to float beside the boat and keep the fish close by. However, a few minutes later a Frigate Bird dived down and stole the fish and the Booby took off after it.

After lunch, we had a walk on Moreno Point, Isabela Island. This was one of the most difficult places to walk because the lava was very bumpy. Extreme care was necessary to avoid a nasty fall. It was like walking on a lunar landscape. It seemed very barren and was extremely hot. However, there is a large lagoon in the centre of the island which, apparently, never dries out but continues to fill with rainwater and groundwater. Common Gallinules (Moorhens), White Cheeked Pintail Ducks and Pink Flamingos had all taken up residence. A Spotless Ladybird landed on my shirt and we saw Lava Lizards as well as Mullet and Skipper fish. It was another amazing walk but we were glad to get back to the boat because it was so hot. Time for a quick shower before we set sail again.

We had an early briefing that night as we had a 13-hour sail ahead of us and the forecast was that it might get a bit rough. We decided to retreat to our room immediately after dinner as the boat had started to rock. It was a challenge trying to get changed into night clothes with the boat rocking so much. We were both in bed by 7 pm. The boat rocked and rolled all night but we did manage to get some sleep. It was a great relief to hear the anchor being dropped at 6 am.

Breakfast next morning was at 7 am and everyone had made it safely through the night. We were due to visit the Charles Darwin Research Centre. By 8 am we were in the pangas and heading for the dock where there were lots of Marine Iguanas waiting to meet us. Walking round the research centre was very interesting and, on the path, up to the centre we saw two new finches – Vegetarian and Cactus. We got to see the hatchling tortoises and managed to photograph some Yellow Warblers before going to the pen where Lonesome George used to live. It is a pity we were a couple of years too late to see him. We did see his girlfriends though. They must have been feeling lonely without him. With George gone it was necessary to find a new attraction. Another Giant Tortoise called Diego turned out to be a good choice. Unlike George, Diego is a super stud and has fathered many babies which have been released back onto the island he came from. We also saw some Land Iguanas which the Centre had been doing research on. However, recently, funding has become a problem and as a result they no longer undertake research on the Iguanas.

After our trip around the centre, we were taken to the shop and of course Una had to spend some money. After all it was a once-in-a-lifetime trip. She bought T-shirts for both of us, a sweatshirt, a shot glass and coffee to take home. Because she spent so much, they gave her a free gym bag to carry it all in! We also got a couple of photos of the two of us standing outside the centre with the sign in the background. Now it was time to head into the town and have a couple of hours to ourselves. On the way we passed another shop which had a T-shirt showing the birds of the Galapagos and of course Una couldn't resist it. We wandered on into the town looking at the shops but not purchasing anything as our boat trip was due to end the following morning and we would be staying here for a few days. Una spotted a jewellery shop and thought we would probably get most of the things we wanted there as it didn't seem to be very expensive. We managed to find a shop selling soft drinks so we sat on the street drinking Inca Cola and orange juice.

We met up with the rest of the group again at 11:30 am and it was back to the boat for lunch. Tanya joined us at our table because some of the group had already done this trip and had decided to go off swimming instead. Tanya's parents live in Scotland and she visits them every year with her two children. However, this year her children will be moving to Scotland to go to school. I am sure she will miss them.

That afternoon we had a drive into the highlands of Santa Cruz Island. We stopped at a farm which manages its grounds to encourage wild tortoises to take up residence. During the breeding season the tortoises migrate to the lowlands to lay their eggs and then they return to the highlands. As it was breeding season most of the tortoises had already begun to migrate but we did see a few, including one rather large one that posed for photos with us all.

At the end of this tour there were two large tortoise shells sitting out and Tanya turned to Bertie and Rosie (the two youngest members of our group) and told them the shells were there for us to play tortoise. Bertie and Rosie thought it was a joke but Tanya was serious so they each crawled into a shell and pretended to be tortoises. It was quite funny watching them and then of course everyone had to have a go including us. The shells were unbelievably heavy. It is no wonder the tortoises move so slowly!

Next, we went for a walk in a lava tube[18]. It was a 350m long tube and it was nice and cool. The exit had caved in at some stage so we had to crawl out. We got a bit muddy but no one minded. Luckily Alison (our life saver) had some wet wipes with her so we all managed to clean our hands. Finally, we went to have a look at a couple of twin craters which are believed to be lava tubes that have caved in at some time in the past. We had a short walk here looking for birds but didn't see anything new. Then it was back on the bus and off to the port to get the boat. We were back on board at 5:45 pm and just had time for a quick shower and some packing before the cocktail reception at 7 pm. Everyone seemed to enjoy the cruise and thanked the captain and crew for making it such a wonderful trip. Dinner was at 7:15 pm and this was followed by the briefing session. We had another early start as we had to be on the island at 6 am to see the sunrise. We all sat around chatting for a while before heading to our room to finish packing. We were looking forward to getting off the boat the following day.

We were up the following morning at 5 am to watch the sun rise over South Plaza Island but we had no luck as it was too cloudy. There were Marine Iguanas, Sealions and Swallow-tailed Gulls on the dock to meet us as well as Sallylightfoot Crabs and Frigate Birds. We had an interesting walk round the island and saw a Cactus Finch doing what they do best – eating the cactus fruit.

---

[18] A **lava tube** is a natural conduit formed by flowing **lava** which moves beneath the hardened surface of a **lava** flow. **Tubes** can drain **lava** from a volcano during an eruption, or can be extinct, meaning the **lava** flow has ceased, and the rock has cooled and left a long cave, or tunnel.

We also saw a couple of dead Iguanas. Apparently there had been no rain for quite some time so they had dehydrated. We saw some individual Iguanas shading under cactus trees and Tanya told us that they have to fight to get a tree and then hold on to it. When we got back to the dock, we saw a baby sealion feeding from its mother. Tanya reckoned it was about three months old. I spotted a baby Swallow-tailed Gull still covered in down so it must have been very young.

Once we were all back on board, we set sail for the airport. We had breakfast and then it was time to finish packing and leave our bags outside the door of the cabin. However, Tanya told us to leave ours inside the cabin as we were not going to the airport but staying on Santa Cruz for a couple of days. We would be leaving the boat with Richard and Jane after the others had been taken to the airport. We had booked into the Finch Bay hotel. Richard and Jane told us they would be staying there also for two days before heading to the Amazon. The boat moored about 8:15 am and we went up to say goodbye to everyone. There were hugs all round, we wished everyone safe onward journeys and then they were off. We weren't leaving until 9:30 am so we joined Richard and Jane, as well as Glenn and Lesley (who were staying on the boat) for coffee in the lounge.

The panga dropped us off at the dock at 9:45 am and a pickup truck was waiting for us. We then had an hour's drive through the highlands (where we had been the day before) to Puerto Ayora. Next, we had to transfer to a water taxi for a five-minute trip over to the hotel landing. This was followed by a 10-minute walk to the hotel. By now it was 11:15 am and it was already very hot. However, our driver had stayed with us and got a barrow to carry our big cases. It must have been hot for him too as he would stop now and then to catch his breath. About halfway along the walk there was a small lake on the right and standing in the middle of it was a Stilt. Another new bird to add to our list. As we passed along the beach, we saw yet another new bird – a Ruddy Turnstone. At reception we were given orange juice and cold towels to freshen up. Check-in time was normally 1 pm but they told us to take seats and they would let us know when our rooms would be ready. Luckily, we only had to wait a few minutes. We had a comfortable room with a veranda overlooking a small pond which was home to a few White-Cheeked Pintail ducks – perfect for us. We unpacked the things we would need for the next couple of days.

Richard and Jane were also having lunch so we joined them. We sat chatting for ages before heading back to our room for a rest. We headed into town at 4 pm and went to the Galapagos Jewellery shop to look for souvenirs. They had some little charms which weren't too expensive so we got a few but decided we needed to make a list so that we would know how many we required. We wandered back up through the town looking at the shops to see what else we might buy tomorrow. Then we caught a water-taxi to take us back to the hotel. We decided to have a quick dip in the pool. It was still very warm and there were some small flies about but we enjoyed the dip. We were joined by a White-cheeked Pintail who seemed to enjoy the pool too. When getting dried off I seemed to have been bitten all over by the flies. They always seem to prefer my blood to Una's! We must remember to bring the bug spray tomorrow. Time for a quick shower and then down for dinner. We had a quick chat with Richard and Jane before heading back to our room.

It was nice not having to set the alarm or get up at a set time next morning. Still, we were the only ones in the breakfast room at 8am. We were joined by finches and mockingbirds which were following an evolutionary route Darwin might not have considered. The finches helped themselves to cereal and the mockingbirds liked the bread. It was great fun watching them. We headed into town at 10 am. We went straight to the jewellery shop and the owner recognised us from the day before. I think he was happy to see us coming as we spent quite a bit. He asked when we were leaving and Una said he would be sorry to know we only had one more day. We got all the things we needed and wandered back up the town looking at the shops. Una spotted some crystal ornaments with "Galapagos" engraved into them – just like many others we have collected on our trips. We had to get one. We were back in our hotel at 12 noon and had some lunch sitting by the pool. After lunch I had a rest while Una caught up on some notes. We went for a swim in the sea at 4 pm. This was more enjoyable than the pool and there were no flies to bite me. Richard and Jane had suggested we meet up for dinner but we hadn't seen them all day. Una said we should sit up at reception and if there was no sign of them by 7 pm we would just eat in the hotel. They arrived back about 6:30 pm and told us they had been at Tortuga Bay all day and it was really worth a visit. We decided that would be our trip for tomorrow. They had booked a table in La Garrapata in town and wondered if we would join them. We met up at 7 pm and took the water taxi back to town. We had a lovely evening chatting to Richard and Jane. They were leaving for the Amazon in the morning so we wished them a safe journey. We were in bed by

10 pm as we wanted to get to Tortuga Bay the following morning before it got too hot.

We were up at 6 am. There were no birds to join us for breakfast as all the shutters were still closed. We headed off just after 7 am and it was already getting hot. We walked up past the hospital and then checked with someone if we were going the right way. It was about a 20-minute walk to the entrance of the park and then we had a further 45-minute walk to the bay. The first part of the walk was along a very long path surrounded by trees and cacti. There wasn't much to see. It was however shaded from the sun which helped a bit. It took us 30 minutes to reach a beautiful beach but swimming isn't allowed there because the currents are too strong. We walked along the beach for a further 15 minutes before rounding a corner and finding ourselves in Tortuga Bay. It really was worth the walk. There were Marine Iguanas basking in the sun and White-Tipped Reef Sharks swimming round our feet. We found a shady spot under the trees to place our stuff before heading off to get photos of the sharks. There were a couple of people paddling with the sharks and a young girl who seemed afraid of them. An older man appeared and tried to catch one of the sharks. It is illegal to touch any of the animals on the islands. He did eventually catch one but when it tried to bite his arm, he let it go. The group left then and once the sand settled down, we managed to get a few photos and some video. I went in for a swim first as we couldn't leave our cameras unattended on the beach. I didn't stay in too long as it was now very hot and I didn't want to get sunburned. The water was beautiful and you could walk out a long way before it got deep. It is one of the most beautiful beaches I have ever seen. After our swim we decided to make our way back to town. We paddled our way along the beach to the path and then changed back into our walking shoes. When we finally got back to the park entrance, we stopped for ice cream to cool down. Una said it was the nicest she had ever tasted. We were back in town at 11:50 am and went to get the last couple of presents we had forgotten about. We were going to have lunch but the place we chose wasn't open. We decided to return to the hotel for a shower and have lunch later. We started packing as we were heading home next day. Luckily all the things we had bought were small and light so there wasn't any difficulty getting them into our luggage. Dinner was scheduled for 7 pm in the hotel. We went up a bit early and sat relaxing at the pool. Our pick-up time in the morning was 7:30 am so it was back to the room to finish packing and get an early night.

We were up next morning at 6 am and down for breakfast at 6:30 am. We packed the last bits and pieces in our luggage and then went down to reception. The temperature had dropped and it was raining a bit which suited us. Our luggage went on ahead of us but our driver hadn't arrived to pick us up. The receptionists tried to phone him but got no answer. He eventually turned up about 15 minutes late and we walked round to the dock. The taxi was waiting for us with our luggage on board. Then we had an hour's drive before boarding a ferry for the short crossing to the island where the airport is. Our driver accompanied us all the way to the airport and helped us check in to make sure we had no problems. We took off about 10:45 am and had a very smooth flight. Breakfast was served just after take-off. We landed in Guayaquil at 12:20 pm and had a very long wait until our next flight. We made our way through to the international side of the airport but our flight wasn't listed as it was only 1 pm and we aren't due to take off until 8:40 pm. There wasn't much to do as it is a very small airport and only had three shops and a cafe. We had lunch at 2 pm. Una had a look in the shops and spotted a beautiful vase decorated with Toucans and Parrots. I liked it too, so we bought it.

We passed the time people-watching. There were a lot of long goodbyes. One boy must have been going away for some time as he was hugged by his mother (I assume) for well over 5 minutes. Then he still had to hug and kiss the rest of his family. It took over 15 minutes and a lot of tears to finally say goodbye. Our flight opened at 5:30 pm and we were able to book the cases right through to Belfast. We made our way to security where they asked to look at Una's rucksack. They had spotted the crystal ornament we had bought on the islands and once they saw what it was, they let us go on our way.

We found the VIP lounge which was right beside our gate and were sitting relaxing, congratulating ourselves that we wouldn't see our cases again until we reached Belfast when I heard my name being called out over the tannoy system. Apparently, the drugs squad had pulled our cases and wanted to check them but could only do so in our presence. We had to bring our passports and wait for them to get to our cases. The only good thing was that they seemed to be pulling most people so I wasn't too concerned. When staying in the Finch Bay Hotel we were left a chocolate on our pillow each night. They always had the head of an animal or bird of the islands on them. We don't eat chocolate so I collected them and decided to bring them home as presents. The drugs officer took a great deal of interest in these. He lifted one out and stuck his pocketknife through it before squashing it between his fingers. Having done that, he decided they were OK and we were allowed to return to the lounge. There was just time for a quick drink before boarding our flight. An interesting end to our trip to Ecuador.

We boarded at 8:25 pm and when we found our seats there was a bag and coat sitting on Una's seat but no one was about. A few minutes later a lady appeared and insisted the seat was hers. One of the cabin crew appeared so I explained the situation to her and showed her our boarding passes. It turned out that the lady was in the wrong seat and had to move further down the plane. She wasn't too happy about it.

We took off at 9 pm about 20 minutes late and dinner was served immediately. I passed the time sorting through my photos while Una was writing postcards which we were taking home as we hadn't been able to get stamps to post them in Ecuador. After a smooth, uneventful flight we landed in Madrid at 2:30 pm local time. We were in the air again at 4 pm for a short two-hour flight back to London. We got the bus to Terminal 1 and made our way through to passport control. We used the self-check-in to pick up our boarding passes and went through to the VIP lounge. It was now 29 hours since we had left the Galapagos Islands and we still had a few more to go. We boarded our last flight at 7:40 pm and took off at 8:30 pm. The sunset was amazing and the colours in the sky were fantastic – reds, blues and oranges. We landed in Belfast at 9:30 pm and, just as we expected, our cases didn't make it. We filled out a lost luggage form and took a taxi home. It was now 31 hours since we set off from the Galapagos Islands. We were very tired but happy to be home.

We were up bright and early next morning and out to Templepatrick to collect Kiki. She remembered us and seemed happy to see us. Rosemary said she had had a ball with the other dogs. She had befriended one of her miniature poodles, Tina, and spent a lot of time following her around. It was good to have her home. She was now three and a half months old. Later in the afternoon one of our cases arrived. It was the one that didn't contain the presents. We checked the website and it appeared that no one had any idea where the other case was. It did eventually turn up the following afternoon and all the presents were intact. So ended our trip of a lifetime and what an amazing trip it was.

France

(1997, 1998, 2000, 2007, 2009)

Strasbourg And the Council of Europe

In talking to people, I find there is often confusion as to what exactly the Council of Europe is. Those with a little knowledge assume I am talking about the European Council. Others assume that I mean the European Union. So, let me begin with some clarifications.

The Council of Europe

The Council of Europe is the continent's leading human rights organisation. Its headquarters is in Strasbourg. Founded in 1949, its stated aim is to uphold human rights, democracy and the rule of law in Europe. It has 47 member states, covering approximately 820 million people. 27 of the 47 states are members of the European Union.

The European Council

The European Council is an institution of the European Union, consisting of the heads of state or government from the 27 member states together with the President of the European Commission, for the purpose of planning Union policy.

The European Union (EU)

The EU currently has 27 members[19] that have delegated some of their sovereignty so that decisions on specific matters of joint interest can be made

---

[19] Following Brexit, 31 January 2020

democratically at European level. No country has ever joined the EU without first belonging to the Council of Europe.

The European Convention on Human Rights

The European Convention on Human Rights is a Council of Europe treaty securing civil and political rights.

The European Court of Human Rights

The European Court of Human Rights, based in Strasbourg, is the only truly judicial organ established by the European Convention on Human Rights. It is composed of one Judge for each State party to the Convention and ensures, in the last instance, that contracting states observe their obligations under the Convention. Its judgments are binding on the countries concerned. The Court has operated on a full-time basis since November 1998.

The International Court of Justice

The International Court of Justice, based in The Hague, is the Judicial body of the United Nations

.

The Court of Justice of The European Union

The Court of Justice of the European Union, based in Luxembourg, ensures compliance with the law in the interpretation and application of the European Treaties.

The Universal Declaration of Human Rights

The Universal Declaration of Human Rights, was adopted by the United Nations in 1948 in order to strengthen the protection of human rights at international level.

The Charter of Fundamental Rights

The Charter of Fundamental Rights is a European Union text on human rights and fundamental freedoms, adopted in 2000.

My first involvement with the Council of Europe was in 1998. The Council of Europe organised a seminar on *Corruption and Organised Crime in States in*

Transition: Juvenile Delinquency and Organised Crime, in Strasbourg, on 27/28 April that year. This was part of "the Octopus Project", a project aimed at assisting states most at risk from organised crime. Northern Ireland was considered to be a "*State in transition*". I was invited as an *Expert/Consultant* to talk about the impact of organised crime on juvenile delinquency. I was also asked to chair a number of sessions.

In the year 2000 I was sent to Kosovo as a Council of Europe expert on the European Convention on Human Rights (ECHR). I was asked to conduct three one-week seminars during March and April to bring the judges up to speed on the ECHR.

(See Kosovo report p135).

In September 2007 I was in Strasbourg for a Council of Europe Conference on *International Justice for Children*. I presented a paper on *The principles of child-friendly justice at national level* which was later published by the Council of Europe. I was back in Strasbourg in October to address members of the "*Lisbon Network*" who were drafting recommendations for the Council of Ministers on *How to Train Judges in Council of Europe Instruments And In Their Implementation.*

In March 2009 I was in Toledo, Spain as a Council of Europe expert working on their project "Building a Europe *for* and *with* Children". The programme comprised two loosely related strands:

- The promotion of children's rights;
- The protection of children from all forms of violence.

The programme's main objective was to help all decision makers and players concerned to design and implement national strategies for the protection of children's rights and the prevention of violence against children. In Toledo we were focussing on "The Protection of Children in European Justice Systems".

The above examples will give you some idea of the kind of work the Council of Europe does and of my involvement with them.

Despite numerous trips to the Council of Europe headquarters over the years I saw little of the city. So, what springs to mind when I think of Strasbourg?

When I hear the word Strasbourg, the thoughts which spring to mind are: alternative home to the European Parliament, home to the Council of Europe and the European Court of Human Rights. I think of efficiency. In my dealings with the Council of Europe I found the staff highly organised, highly efficient and very friendly. Everything ran like clockwork. It was a pleasure to work with them.

Strasbourg is truly at the heart of Europe's political and economic life. It stands in the centre of the Rhine valley, which, for millennia, has acted as a principal trade route between Northern and Southern Europe. Strasbourg and the Alsace region are uniquely situated for business, technology, trade, and transport.

It seems strange to say then that I found it a difficult place to get to. And I am not the only one. I read a report by Joe Roberts in the Metro News (2 July, 2019) that newly elected Brexit Party MEP, David Bull (better known as a TV doctor) has been ridiculed for complaining about the 'insanity' of his new commute to work. He moaned that it would take him eight hours to travel all the way from Ipswich to the French city near the German border. He said, "For some reason the parliament seems to be in a very inaccessible place." I have to admit that I felt the same way about the headquarters of the Council of Europe. Bull complains about having to take four trains. On one occasion I had to go by plane, train, bus and taxi to get from Belfast to Strasbourg on time. But, for me, this was a minor irritant, not a major issue.

The Cathedral of Our Lady of Strasbourg stands on an island formed by the canalization of the River Ill in the centre of the city. At 142 metres tall, it dominates the landscape. I understand that it is visible across the flat Alsace River plain from as far away as the Vosges Mountains or the Black Forest in neighbouring Germany. Considered to be one of the finest examples of Late Gothic architecture in Europe, the Cathedral is also the sixth-tallest church in the world. Unfortunately, while I saw it from a distance, I never had an opportunity to visit it.

Something else I couldn't miss seeing, because of its size, was the white stork. It is a large bird – 100-115 cm (39-45 in) long and stands 100-125 cm (39-49 in) tall. The wingspan is 155-215 cm (61-85 in). Males are larger than the females and can weigh up to 4.5 kg. They have long red legs, a long neck and a long straight pointed beak which is also red. The plumage is mainly white with black flight feathers and wing coverts. They are large, spectacular birds. It would be difficult to miss them standing on the rooftops. It is estimated that there are about 50 breeding pairs in Strasbourg.

Coming from Ireland, renowned for large families when I was growing up, readers might find it hard to believe that I had never seen a stork. You would imagine that it would be difficult to miss such large birds, carrying new-born babies wrapped in a blanket in their beak, crisscrossing the parish on the way to some neighbour's house – there always seemed to be at least one house in the parish with a new-born baby.

Seriously, readers will be aware that I am a keen birdwatcher and never go anywhere without my binoculars. There have been reports of rare sightings of storks in Ireland (July 9, 2016 in Cork, May 12, 2011 in Derry, for example). But I had never seen one, not even on my many trips around Europe. And storks were not on my mind as I walked up to my hotel on my first visit to Strasbourg. Consider my delight on spotting several storks standing on nearby rooftops. Now, when I think of Strasbourg, that image of the storks springs to mind.

India
2005

In December 2005 I was in Lucknow, India, to attend "The 6th International Conference of Chief Justices of the World entitled *Towards a New World Order*". I was invited to attend as President of the IAYFJM. I was asked to present a paper to the conference on "The Child's Right to Education". The title seemed very appropriate in view of the venue.

The conference was held in the City Montessori School in Lucknow. Founded by Jagdish Gandhi and his wife Bharti with a loan of just 300 rupees, the school opened in 1959 with five pupils. The Guinness Book of World Records, 2013 edition, states that the school had a world record enrolment of 39,437 children on 9 August 2010 for the 2010-11 academic year. The school was still growing. 47,000 pupils were enrolled for the academic year 2012-13. The school admits boys and girls between ages two and five, who can then continue their education to degree level.

I didn't see much of Lucknow. I didn't even see much of the school! The school was so big it was like a town. As I walked from my accommodation to the venue and back all I could see around me was school buildings.

From Lucknow I travelled back to Delhi where I planned to stay for a couple of days before heading home. Delhi is a city that bridges two different worlds. To that extent, it reflects India itself. Old Delhi, once the capital of Islamic India, is a labyrinth of narrow lanes lined with crumbling "havelis" – historic and architecturally important private residences – and formidable mosques. In contrast, the imperial city of New Delhi created by the British Raj is composed of spacious, treelined avenues and imposing government buildings. New Delhi is the capital of the Republic of India, and the seat of executive, legislative, and judiciary branches of Government.

If there was one thing that intrigued me most about India in general and Delhi in particular it was how Hinduism's Sacred Animal – the Holy Cow – could co-exist with heavy vehicular traffic in modern cities.

I was aware that respect for animals has been a central theme in Hindu life for more than 3,000 years. I share their respect for animal life but wondered how wise it was to allow cows to wander freely through streets crowded with vehicular traffic. In 2003, Bibek Debroy, a columnist for India's Financial Express, wrote: "What is the greatest traffic hazard in Delhi today?" "Cows." Two years later, I would have had the same response. I don't know what the city's population was in 2005 and had no way of knowing how many cows roamed the streets. But as cities have grown more crowded, cow-friendly policies have posed problems. In 2008 it was reported that Delhi's 13 million residents shared the streets with an estimated 40,000 cows. A number of initiatives have been tried over the years but control is difficult, bearing in mind that cows cannot be harmed.

A number of Indian states have recently created new policies around bovines. In autumn 2018, Madhya Pradesh announced the creation of a cow ministry. On January 10, 2019, the Hindustan Times reported that Delhi's development minister Gopal Rai plans to let cows and elderly people coexist in a shared space where they can both enjoy their golden years together. Rai plans to open *"the most advanced cow shelter"* in an area that would also be a home for humans, *"where the elderly will be in the company of cows. An 18-acre plot will be allocated for this"*. With Delhi's population currently at 26 million, and still growing, clearly something needs to be done. But, as I approach my 82nd birthday, I am not sure that I would appreciate the thought of being put out to pasture!

New Delhi is not too far from Agra and the Taj Mahal[20]. I couldn't leave without seeing that. I feel there is no need to describe the Taj Mahal because it is so well known. Built entirely of white marble it is regarded as one of the eight wonders of the world. UNESCO described it as the jewel of Muslim art in India and one of the universally admired masterpieces of the world's heritage. No image can do justice to the intricacy of the detailing or the subtle hues and ever-changing colour of the marble. Suffice to say that I agree with those who believe

---

[20] It was commissioned in 1632 by the Mughal emperor Shah Jahan (reigned from 1628 to 1658) and completed in 1653.

that its stunning architectural beauty is beyond adequate description. Nothing compares to seeing it "in the flesh". It was a wonderful experience to be there.

Before leaving Delhi, I wanted to see the Red Fort[21], a symbol of India, dating from the Mughal[22]-era. I took a three-wheel taxi[23] to Old Delhi.

The Red Fort was the main residence of the Mughal emperors for nearly 200 years. It is one of the building complexes of India encapsulating a long period of history and its arts. Two years after my visit it was designated a UNESCO World Heritage Site. The location is currently used by the Prime Minister of India to address the nation on Indian Independence Day.

It has an area of 254.67 acres enclosed by 2.41 kilometres (1.50 miles) of defensive walls, punctuated by turrets and bastions and varying in height from 18 metres (59 ft) on the river side to 33 metres (108 ft) on the city side. The marble, floral decorations and double domes in the fort's buildings exemplify later Mughal architecture. Its artwork is a blend of Persian, European and Indian art, resulting in a unique Shahjahani style rich in form, expression and colour.

The fort stands on the banks of the Yamuna River, which once fed the moats surrounding most of the walls. The Yamuna is the only major river flowing through Delhi.

As I approached the Red Fort, I noted a crowd of people on steps leading down to the water's edge. Some people were bathing in the waters. Others were in attendance at three or four cremations.

I was aware that, in Hinduism, the river Ganges is considered sacred and that Hindus believe that bathing in the river brings the remission of sins and facilitates Moksha (liberation of the soul from the cycle of life and death). Cremation on the banks of the Ganges frees the soul from reincarnation.

I had never heard of the Yamuna and did not know that it is one of the country's most-sacred rivers. As I watched the people bathing, I noted that the water was extremely polluted. My driver told me that those who enter the water emerge caked in dark, glutinous sludge.

In the 16th century, Babur, the first Mughal emperor (1526-30), described the waters of the Yamuna as "better than nectar". For the first 250 miles (400 km)

---

[21] Emperor Shah Jahan commissioned construction of the Red Fort on 12 May 1638, when he decided to shift his capital from Agra to Delhi.

[22] The Mughal era: Muslim dynasty of Turkic-Mongol origin that ruled most of northern India from the early 16th to the mid-18th century.

[23] Called a "tuk tuk" in the Philippines but, simply, a "three-wheeler" here.

from its source in the lower Himalayas, the river still glistens blue and teems with life. And then it reaches Delhi. In India's crowded capital, the water is siphoned off for human and industrial use, and replenished with toxic chemicals and sewage from more than 20 drains. No wonder the river looked polluted.

Legend recounts how the gods and demons fought over the pot containing the elixir of immortality. During the struggle, drops of the elixir fell on the Kumbh Mela's[24] four earthly sites, one of which was here on the Yamuna. The rivers are believed to turn back into that primordial nectar at the climactic moment of each, giving pilgrims the chance to bathe in the essence of purity, auspiciousness, and immortality. Clearly, for the pilgrims, the only thing that mattered was that, as they bathed, their sins were being washed away and their soul released from the cycle of life and death. The polluted sludge could be washed off later.

My driver pointed out a nearby hotel and told me that all the residents were old people who wanted to devote their remaining years to a period of reflection, meditation and prayer. Here, the entire culture is geared towards prayer. Mornings allow time for peaceful reflection, in the presence of likeminded holy people, down at the ghat (steps leading down to the water's edge). The afternoon *satsang* (listening to holy scripture or music), allows the elevation of one's mind from worldly concerns towards a higher level of thought. Each evening there is a religious ritual of worship (*artis*) in which light (usually from a flame) is offered to one or more deities.

I took one last look at the funeral pyres, reminding myself that Hindus speak of death as the Great Departure, regarding it as life's most exalted moment. The death anniversary is called Liberation Day.

As I got into the taxi to return to my hotel, pick up my luggage and head for the airport, I reflected that I had come here intent on learning a little about India's history and art. I left having, unexpectedly, gained a valuable insight into its people and Hinduism.

---

[24] Kumbh Mela is a major pilgrimage and festival in Hinduism. It is celebrated in a cycle of approximately 12 years at four riverbank pilgrimage sites.

Iran
2006

Tehran

I was surprised to receive an invitation to conduct a judicial training programme in Tehran, Iran in 2006. The UN had imposed sanctions on Iran and the Americans were energetic in their efforts to make sure the sanctions were observed. I was working for UNICEF (The United Nations Children's Fund) who have their headquarters in New York. But ours not to wonder why...! My role was to conduct a *Training of Trainers* programme "for judges and suitable law faculty lecturers on the Convention on the Rights of the Child, juvenile justice and best practices". The mission would run from December 4 to December 11.

The judges would have guessed that I was not an expert in Islamic law so it was important that I had my homework done. I was able to tell them that I was aware that the head of the judiciary, Ayatollah Mahmoud Hashemi Shahroudi, had issued a circular in 2003 in which he requested judges not to issue execution verdicts for children under eighteen. I understood that this was a request and not an order. It was unlikely to end juvenile executions, as many judges would not listen. They could argue that they had to abide by the law as laid down in the Penal Code. A simple request, even from the head of the judiciary, could not override the legislation.

The members of my group insisted that the majority of judges were opposed to the use of the death penalty in general and the use of stoning to death in particular. They believed that pressure from human rights groups was having an impact and they looked forward to it being discontinued altogether.

The age of criminal responsibility was another controversial issue. According to Islamic sources, the criterion for criminal responsibility is reaching the age of maturity which, in Iran, is related to the age of puberty. Under Iranian law the age of criminal responsibility is set at 14 years and 7 months for boys and 8 years and 9 months for girls, which is not only low by international standards but also discriminatory.

If a boy who is not yet 14 years and 7 months old was charged with murder or a lesser crime like illicit sexual behaviour, he would be subject to minor correctional and security measures such as submission to his parents with a promise of correction, sending the child to a social worker or psychologist, banning him from visiting specific persons or places. If a girl aged 8 years and 9 months or older is charged with murder or lesser crimes like illicit sexual relationships, she could be sentenced to violent and inhumane punishments such as flogging, amputation of the right hand and left foot or even stoning to death. A girl seen in the company of a member of the opposite sex other than her immediate family – even if she is just talking to the boy next door – may be charged with having an illicit sexual relationship.

I pointed out that internationally accepted standards have determined that the age of 18 is the standard age of entering into majority and full criminal responsibility, without any discrimination between boy and girl.

In discussing how to resolve the problem I pointed out that, in drawing up their Penal Code (as it existed at that time), there had been many legal and religious disagreements about the age of maturity and criminal responsibility. Some Islamic jurists held different views on the age of maturity – for example some proposed the age of 13 for maturity of girls. In addition, the majority of lawyers believed that recognition of criminal responsibility for a girl of 8 years and 9 months old and a boy of 14 years and 7 months old was wrong and violated international standards including the Convention on the Rights of the Child.

I suggested that it was unlikely that the Iranian authorities would accept raising the age of maturity for boys and girls to 18 or even bringing the age of maturity for girls into line with that for boys. I proposed that we as a group recommend to the authorities that the age of maturity for girls be raised to 13 as an interim step. The group agreed unanimously.

I heard afterwards that our recommendation was rejected by the Iranian parliament at that time. A new Penal Code was adopted in 2012. The law still maintains the age of criminal responsibility at 8 years and 9 months for girls and 14 years and 7 months for boys. The new law makes some important steps forward but it does not completely abolish juvenile executions and falls short of requirements under international law.

The dual standards over sanctions were highlighted for me before I left Iran. UNICEF's country rep apologised for not being able to pay my expenses – they were endeavouring to sort things out. UNICEF New York was unable to send the money to UNICEF Tehran because the sanctions did not allow it. They worked out a compromise. UNICEF New York transferred the money to their office in Geneva. The Geneva office then transferred the money to the Tehran office. If UNICEF could get around the sanctions so easily, I was sure others could too.

I had no time for sightseeing in Iran. I saw very little of Tehran, apart from the short commute from my hotel to my place of work. A car was sent to pick me up in the morning and return me to the hotel in the evening. My driver's English was quite good but the sole topic of conversation was the traffic which was always at a virtual standstill during rush hour. And, of course, there are always those drivers trying to cut across to the "fast" lane only to discover that the lane they just left is now moving faster than any other. So, back they go again.

One thing I did notice was that there were lots of shops selling carpets. This was hardly surprising since Iran is the world's largest producer and exporter of carpets, producing ¾ of the world's total output. When I was asked to go to Iran, I decided that I couldn't come back without a Persian rug. So, each evening, when I got to my hotel, I would go out walking and visit at least one carpet shop.

At that time of the evening the traffic was still gridlocked and it was difficult to see where you could cross the street. Some locals, noting my hesitancy, called me to follow them as they weaved through the traffic. This was typical of all the people I met – relaxed, friendly, always ready to lend a helping hand.

The people in the carpet shops were equally friendly and helpful. But they were hoping for a big sale. I saw some beautiful carpets and was offered fantastic discounts. When I explained that I wanted a rug measuring 40x26 they wanted to know if that was metres or yards. When I replied "inches" they thought I was joking. I said I wanted to be able to leave the shop with the rolled-up rug under my arm and place it in the overhead locker on the flight home. They insisted that size was not an issue. They were prepared to ship the carpet to Belfast at no extra cost to me. What none of the shop keepers could do was to give me what I wanted. They didn't have a rug that small!

On my final day I asked my driver if he had any suggestions. He had. That evening he took me to a carpet museum where I saw some of the most beautiful carpets I have ever seen. Afterwards he took me to a shop where I was able to buy a rug to suit my requirements. So, I have my Persian rug to remind me of a country and a people very different to the media portrayal.

Italy
1990, 1996, 1997

Turin

I first visited Italy in 1990 to attend the IAYFJM World Congress which, on that occasion, was held in Turin. I didn't know a lot about Turin other than it was an industrial city where some of Italy's best-known cars (Fiat, Lancia and Alfa Romeo) are manufactured; it was the home of the Turin Shroud and had two football teams – Juventus and Torino.

My most vivid memory of Turin is its covered walkways. When Turin was the seat of the Savoy kingdom, the kings and queens enjoyed strolling from their palace in Piazza Castello, down to the River Po so much, that they had covered walkways built all the way through the city centre. They could enjoy their walk all year round, whatever the weather.

These arched walkways are a beautiful feature of today's city centre. I understand that there are 18 km of arcades, of which 12.5 are interconnected. This gives the city the largest pedestrian area in Europe. The arcades offer protection from summer heat or winter snow, or rain in my case. Walking under the arcades it is easy to be enchanted by the beauty of paved, coffered ceilings, rows of columns, and glimpses of the halls of historic buildings.

I had hoped to see the Turin Shroud but I found exceptionally long queues. I was advised that waiting time was likely to be several hours. Apparently, the Shroud had been made available for viewing after a long absence. Hence the long queue. I decided to continue exploring the covered walkways.

Turin is also famous for its coffee and chocolate. The Italians make their coffee much too strong for my taste and I don't like chocolate. So, there is no way I was going to try a *Bicerin*. This is a drink unique to Turin made with layers of hot chocolate, coffee and cream. But I did buy some bags of *giandujotti*. These are made from a mixture of hazelnuts and chocolate and I thought they would go down well with folk back home.

Palermo

In April 1996, I attended a conference in Palermo. My outstanding memory of the conference was the high level of security. We weren't told officially but heard on the grapevine that a local judge had been murdered recently for not taking directions from the Mafia. There wasn't a police officer to be seen in the conference hall but there were a number of muscular young men standing around the room, backs against the wall and whispering up their sleeve from time to time. And I thought they only did that in the movies!

There was a lot to see in Sicily but I had very little time to see it. I wanted to climb Mount Etna. I took a bus to Catania intending to stop over and go on to Mount Etna the following morning. Consider my disappointment to be told that Etna was out of bounds because it had been spewing out volcanic ash for the past couple of days. Still, I enjoyed the bus trip across the island.

Naples

The Executive Committee held a meeting in Naples, in April 1997. Naples is situated in a beautiful part of Italy but blighted by highly organised criminal gangs directed by the local mafia. There were pickpockets everywhere. One of our group was mugged while withdrawing money from a cash machine. I had a bottle of water taken from my backpack. I am not sure if the thief was thirsty, it was a hot day, or if it was just a practice run. I went to the public toilets to find them full of druggies many of them openly injecting themselves. I never expected that I would be glad to get out of Naples. I took a daytrip to Capri.

Capri was, for me, a mythical place, a place made known to the world in the writings and legends of Ancient Greece. The island was first colonised by the Greeks and was later adopted as a possession of Naples. The Emperor Augustus visited it in 29 BC and was so taken by its beauty that he traded Ischia for it with the city of Naples.

The crags and grottoes of Capri have been enchanting visitors ever since. Set in the bluest ocean you've ever seen, its breath-taking landscapes and dramatic views, with the scent of lemon blossoms in the air, has been an inspiration to artists, poets, lovers and travellers throughout the centuries. Johann Wolfgang von Goethe (a German writer and statesman), for example, referred to Italy in his poetry as the land where the lemons bloom.

Capri had long been on the list of places I would like to visit. I would never get a better opportunity. It was just a few minutes' boat-ride away. There wouldn't be time to visit the many tourist attractions but all I wanted was a couple of hours to chill out, have a coffee and enjoy the view. In Homer's epic poem, Odysseus almost succumbed to the voices of the Sirens as he sailed past the island. My visit would be so brief that there was little likelihood of me falling under their spell.

Pompeii

Another place I always wanted to visit was Pompeii. Once a thriving and sophisticated Roman city, Pompeii was buried under metres of ash and pumice after the catastrophic eruption of Mount Vesuvius in 79 AD. Vesuvius erupts on a regular basis. The eruptions are generally preceded by small earthquakes. The Roman writer, Pliny the Younger, wrote an eyewitness account of the eruption and its aftermath. He reported that tremors were felt four days before the 79 AD eruption, gradually increasing in frequency. But they were not particularly alarming because the Romans had grown so accustomed to them. They paid no attention. Pliny the Younger's uncle, Pliny the Elder, succumbed to volcanic gas after wandering too close.

The eruption, when it came, had an explosive force estimated to be 100,000 times more powerful than the Hiroshima-Nagasaki bombings. It ejected a cloud of stones, ashes and volcanic gases to a height of 33 km (21 miles) into the sky. Ash and pumice rained down on Pompeii, blanketing the city and surrounding area to a depth of several metres. Sometime early next morning the first of four pyroclastic-surges swept down the mountain. The flows, travelling at over 100 miles per hour, dense and very hot, destroyed anything that lay in their path, incinerating or suffocating anyone still in the area. The cities of Pompeii, Herculaneum, Oplontis and Strabiae, as well as several other settlements were destroyed. More than 1000 people died – the exact number will never be known.

Encyclopaedia Britannica describes Pompeii now as "a vast archaeological site in southern Italy's Campania region, near the coast of the Bay of Naples. The preserved site features excavated ruins of streets and houses that visitors can freely explore." It attracts 3.5 million tourists a year (or did before Covid-19!)

My sister Una accompanied me on one of my visits to Rome and we decided to take a day trip to Pompeii and explore it for ourselves.

Top of our list was The House of the Faun – the largest and most expensive residence in ancient Pompeii, and the most visited of all the houses in the famous ruins. This house was the residence of a very wealthy family. It took up a whole city block, with an interior of some 3,000 square metres (nearly 32,300 square feet). Built in the late second century BC, the house is remarkable for the lavish mosaics which covered the floors, some of which are still in place.

Next on our list was The House of Vettii. It was described as a Roman townhouse – one of the largest in Pompeii. We were told it was typical of the homes of the prosperous merchant class. The house is divided into two areas: the part in which the family lived, laid out around the main atrium and the peristyle with its beautiful garden; and the part where the servants lived and worked. This luxurious residence houses an impressive collection of fresco decorations typical of the wall paintings in the houses of rich Pompeian traders. There are twelve surviving panels, which depict mythological scenes. There are many more artworks on display. The most memorable is a painting on the wall facing the entrance. The painting depicts Priapus, the god of fertility, weighing his phallic member on a set of scales.

Time to move on.

1,044 casts have been made from impressions of bodies recovered in and around Pompeii, Thirty-eight percent were found in the ash-fall deposits and 62% in the pyroclastic surge deposits. We went to see some of the casts which were on display. The best known is probably The Resin Lady – she captures the horror of the death. Other well-known ones are: a couple who died in a loving embrace; a slave still in shackles. But most loved is the toddler of Pompeii. He was found close to two adults, presumed to be his parents, and another child, presumed to be a brother or sister. All four have their arms raised in what has been termed the boxer pose – presumably a last gasp for air.

Encyclopaedia Britannica recommended exploring the streets. This seemed like a good time to begin.

It was a strange feeling walking along Pompeii's main shopping street – "Abundance Street" – and realising that "shopping obsession/retail therapy" was not a new phenomenon. These were not market stalls to be set up and taken down as required. These were 58 purpose-built shops, effectively a shopping mall. The shutters were raised in the morning and lowered at night. And this was about 200 BC!

The shops sold all kinds of everything including pots and pans, cushions and other textiles, bread, fruit and vegetables, wine and olives (it is said that the Romans drank a bottle of wine a day). There were barber's shops. There were rooftop cafes. I discovered that "Fast food – convenience food – lunch on the go" was not a new phenomenon either. There were shops where you could buy fast food snacks or takeaways. And there was crime. The shop counters had holes drilled in them which were an exact fit for coins currently in circulation in order to prevent forgeries. Like all cities, crime was commonplace. With no street lighting and no police force it was dangerous to walk the narrow streets alone after dark.

Abundance Street itself was a good example of Roman road-building skills and we had seen examples of the many famous fountains when we visited The House of the Faun and The House of Vettii. Unfortunately, we didn't have time to see the baths which were once supplied with hot water, as hot as any jacuzzi. I was pleased we got to see The Forum where everyone, in Roman times, came to hear the news or the gossip, to seal business deals, to talk politics. We did have an opportunity to see The Amphitheatre – an open-air stadium which could accommodate 20,000 spectators. It is another fine example of Roman engineering skills. But I didn't like what it stood for. I couldn't support 20,000 people coming to watch wild animals fighting to the death; criminals being torn apart by wild animals; gladiators slaughter one another – all in the name of sport.

This reminded me that 40% of the population of Pompeii were slaves. The Gladiators were slaves. The girls in the many brothels were slaves. All the manual work in the city was done by slaves. This was the dark side of the Roman Empire. Slaves were the engine which drove it. It is said that 200 million people were enslaved during the time the Romans ruled the known world.

Before leaving, I thought I would like to try the Stepping Stones to cross the street. They look a bit incongruous nowadays but we have to think of the street as it was pre-AD 79, not as it is today. Back then the street would have been little more than an open sewer. Nobody would want to step off the footpath into that. So, the steppingstones were necessary.

As we headed for our train, I thought about an article I read in National Geographic. In summary, the article said that Vesuvius has erupted many times before and since AD 79. It is the only volcano on the European mainland to have erupted within the last hundred years. It is regarded as one of the most dangerous volcanoes in the world because the area around it is the most densely populated volcanic region in the world. 3,000,000 people live near enough to be affected by a major eruption, 600,000 live in the danger zone. The title of the article: Vesuvius – Asleep for now.

Japan
2001

I left Belfast on December 13, 2001 en route for Yokohama where I was to attend the second World Congress Against the Commercial Sexual Exploitation of Children from 17 to 20 December. It seemed fitting that, on take-off, we flew north in the direction of Stockholm, seat of the first World Congress, which I attended in 1996. We flew northwest following the border between Norway and Sweden, crossed directly over Ivalo and out over the Barents Sea. We turned east and flew over Northern Siberia, gradually swinging southeast to cross the Russian coast again east of Khabarovsk and flew south over the Sea of Japan towards Tokyo and Yokohama.

Japan lies on the Pacific Rim at the edge of Asia. It comprises four main islands but has more than 3,900 smaller islands running from Northeast to Southwest in an archipelago which would stretch all the way from Montreal to Miami.

The climate correspondingly varies from the snowy northern tip of Hokkaido to the subtropical region of southern Kyushu and Okinawa. Tokyo, in the middle, has a temperate climate with cold winters and hot summers.

Japan's total land area is greater than that of Germany. In physical size it is the 62nd largest country in the world but in terms of population it is the 11th largest. A range of high mountains and dense forests means that only two fifths of the country is suitable for habitation and farming. With a population of 126.5 million crowded into narrow coastal plains and river valleys, it is the most densely populated country in the world.

The two principal religions in Japan are Shinto and Buddhism. Smaller religions account for an estimated 1,400,000 believers. These are mainly Protestants, Catholics, Greek and Russian Orthodox, Muslims and Jews.

The indigenous religion of Japan was Shinto. One of the major tenets of Shinto is that the imperial family is directly descended from the Sun God, resulting in the Divinity of the Emperor. Divinity was renounced after World War II, but the emperor remains the titular head of the Shinto religion.

The emperor is also the titular Head of State. The country is governed by a parliamentary democracy with a prime minister and a cabinet. The Parliament, or 'Diet' comprises a House of Representatives with 511 seats and a House of Counsellors with 252 seats. The country is divided into 47 prefectures, each with a governor.

Shinto means "the way of the Gods" and has a strong component of nature worship with shrines in places of great natural beauty such as mountaintops and forests. Divine spirits are believed to inhabit waterfalls, unusual rocks, great trees and the like.

Buddhism is now the predominant religion. It was introduced from China about the beginning of the 7th Century and the two religions were fully integrated by the 12th Century. Buddhist and Shinto shrines frequently stand side by side and I noted people coming out of one shrine and going directly into the other to pray there also.

Japan's mountain range is effectively a long string of volcanoes and its many hot springs attract not only the tourists but the natives as well. Unfortunately, this particular tourist had no time to enjoy the luxury of 'taking the waters'.

Tokyo is Japan's capital and largest city with a population of about 13.9 million. It suffered almost total destruction twice in the 20th Century. An earthquake, and subsequent fire, in 1923 destroyed almost all of the old city and killed over 140,000 people. The city was rebuilt without any comprehensive urban plan and became a city of sub-centres and neighbourhoods, even villages, each with their own distinct personality.

The US firebombing of Tokyo devastated the city again in 1945 and killed more people than the two atomic bombs put together.

Unlike the great cities of Europe there is no prevailing style of architecture and no real centre to the city. However, unlike European cities, Tokyo is clean. I saw no rubbish in the streets and saw no one throwing anything, not even a cigarette butt, away. Indeed, I watched people carefully opening sweets, bars of chocolate and packets of cigarettes and then roll the wrappings up and put them into the nearest bin or into their shopping basket, handbag or pocket.

The only other city I have visited which was equally clean was Singapore. But Singapore is clean because harsh penalties are imposed on anyone found scattering litter. I felt that the Japanese keep their cities clean because it is part of their culture.

The cars, lorries and vans were sparkling clean, in showroom condition. I saw a lot of traffic but saw no dirty vehicles. No scope here for graffiti artists who, at home, scrawl "Please wash me" on the sides of vehicles. Trains were equally clean. I didn't use the buses so I can't comment on them.

The trains were extremely punctual. They ran precisely on time. I found them easy to use because the train stations had the names written in English as well as in Japanese. Some of the trains had electronic notice boards at the end of the carriages which announced what the next station would be. Unfortunately, these were not available in all of the trains. Nonetheless I used the trains quite a bit and had no difficulty at all.

Even the air was clean with no noticeable pollution. Despite the heavy traffic there was no acrid smell of fumes in the air. This was particularly noticeable after my recent visit to Kathmandu.

The politeness of the people was reflected in the driving – no cacophony of horns here, no screaming breaks or squealing tires as cars roar away from the traffic lights. Traffic was orderly with everyone patiently waiting their turn.

If there is a centre to Tokyo it is the Imperial Palace. Built on the site of Edo Castle (Edo, meaning 'estuary' was the original name of Tokyo) this is where the Tokugawa shogunate ruled Japan for 265 years. In modem times it is the home of the emperor. Once the largest system of fortifications in the world, the palace was almost completely destroyed in WWII and then rebuilt in ferroconcrete. However, the architectural style, or lack of it, is unlikely to offend one's sensibilities since no one is allowed within sight of the palace. The nearest I got was a view of the outside of the East gate when I visited the Imperial Park. Had I been able to extend my visit for a couple of days I might have had more luck as the public are admitted to the grounds on two days per year – January 2 and December 23 (I left for home on December 20). However, I am not sure that I would have wanted to be in the melee of Japanese nationals all eager to see their Royal Family waving from one of the balconies.

The Imperial Palace's east garden is open to the public every day except Monday and Friday. I was there on Sunday morning but hadn't time to go in. The garden is famous for its ponds and waterfalls but, most of all, for its azaleas.

It was the wrong time of year for azaleas so I wasn't too disappointed. I made do with a quick walk through the outer park with its stunted pine trees and the moat, which surrounds the castle ramparts.

Ginza is Tokyo's best known shopping district. I found a beautiful jacket there, something I had been looking for, for some time. My search had extended to many cities and now here was exactly what I wanted. A quick calculation converted the price from Yen giving me a price for the jacket of more than £400 sterling (that was 19 years ago). Suddenly my concern for others took over. What if someone else was looking for this precise jacket who needed it more than I did? I decided to leave it for someone whose need was greater, and wallet fatter, than mine.

The price tags in Nakamise Street are less intimidating. Unfortunately, the focus there is on kimonos and I couldn't really picture myself in a kimono. Colourful souvenir shops line both sides of the street, so I was able to find some really nice souvenirs.

Spanning the entrance to the street is Kaminarimon or Thunder God Gate. This street leads directly to the main entrance to the Sensoji Temple, better known as the Asakusa Kannon Temple. This is the oldest and most popular temple in Tokyo. According to legend the temple houses a small statue of the Buddhist Goddess of Mercy, found in the Sumida River by two local fishermen in 628. The entire area, including the Temple, was firebombed in 1945. It was eventually rebuilt in 1958 when funding became available.

At the top of the street is an impressive two-storey gate called Hozomon. Passing under the archway you come to a huge bronze incenses burner. When I visited, there were crowds of people milling around 'bathing' in the smoke. Some were wafting the smoke over their face, or head or to various parts of the body. Some were bathing babies in the smoke. This observance is said to ensure a year's good health and good luck to the faithful. The reason for wafting the smoke towards various parts of the body is to cure those parts which are ailing. The only smoke I got was that wafted in my direction by the many hands as I took photographs.

To the left of the arcade is the famous Five-Story Pagoda. Across the courtyard is the main hall of Sensoji. The style is very different from the temples I have seen in Nepal or Myanmar. But then Buddhism in Japan belongs to a different school. Essentially there are two schools of Buddhism. Theravada, or the "Doctrine of the Elders", follows closely the original teaching of Buddha.

Mahayana has evolved closer to Christianity. The Mahayana School is followed in China, Korea and Japan.

Immediately to the right of the courtyard is the Asakusa Jinja, a Shinto shrine. As I indicated earlier, Buddhism and Shinto get along quite peacefully in Japan.

Most visitors to Tokyo visit Tokyo Tower. The Tower opened in 1958 and was the world's tallest self-supporting iron structure. It stands at 333 m tall while the Eiffel Tower is 320m. Despite being taller the tower weighs 4,000 tons against 7,000 tons in the Eiffel Tower. This is mainly a reflection of the improved quality of steel.

There is a special Observation Tower at 250 m but, unfortunately, it wasn't open during my visit. I had to settle for the Main Observatory at 150 m. However, it still provided an excellent view of Tokyo and of Mt Fuji in the distance.

Japan in general and Tokyo in particular is famous for cultured pearls. I visited a pearl gallery to learn how cultured pearls are manufactured. I participated in a draw for a cultured pearl but didn't win it. Perhaps I should have bathed in the incense!

During my brief visit to Japan, I only had time to visit Tokyo and Yokohama. Yokohama, situated some 20 km southwest of Tokyo, is Japan's second largest city, with a population of around 3.75 million. It is a major international port. Like Tokyo it was severely damaged by the earthquake in 1923 when 20,000 people died and 60,000 homes were destroyed. Half of the city was levelled again by American bombers in 1945. The harbour was quickly restored during the Korean war and Yokohama is now one of the busiest and most important trading ports in the world. It is a more cosmopolitan city than Tokyo and many people who work in Tokyo prefer to live in Yokohama and commute daily. The journey takes less than 30 minutes by train.

My hotel in Yokohama overlooked the South Pier with the waterfront to the left. The area reminded me of Baltimore Harbour. Here in Yokohama the Minato Mirai 21 Project was launched in the mid-1980's to turn a huge tract of neglected waterfront into a 'model city of the future', integrating business, exhibition and leisure facilities. The centre piece of the project is the 70-storey Landmark Tower—Yokohama's tallest building. Around the Landmark Tower is Queen Square, a series of shopping malls and including two hotels, one of which was the Pan Pacific where I was staying. At 25 storeys high it was dwarfed by the Landmark Tower. From my balcony on the 17th floor, I could see the beautiful

'Swan of the Pacific', a three-masted sailing ship now used as a training vessel, anchored at the South Pier. I could also see a huge Ferris wheel standing in the centre of a funfair.

The rear entrance to the Hotel opened out onto the Queen Square with its shopping facilities, restaurants etc. A few minutes' walk over a partly-covered bridge, which spanned the main road, brought me to the Exhibition Centre where our conference was held. This is one of the biggest Exhibition Centres in the world. The main auditorium can seat 5,000 people. Since we had a 'mere' 3,334 delegates attending our conference there was some spare capacity. The Exhibition Hall is 20,000 sq. metres. The National Convention Centre, Exhibition Hall and an hotel have been integrated into a unique shape with the theme of "Waves, Wind and Sunlight" which has become the new symbol of Yokohama.

On the waterfront itself is the Nippon Maru Memorial Park. On my first afternoon I checked into the hotel at around 3:00 pm. I was feeling exhausted, not having slept the night before, and was tempted to lie down. However, I wanted to stick to my usual routine of adapting as quickly as possible to local time. So, I determined not to go to bed until night-time. I had noted on my map that the Nippon Maru Park was close by (the word 'Memorial' was not included on the map). I took my binoculars and went looking for the park. I found it without too much difficulty and was disappointed when I discovered it was a memorial park. There were very few trees and, consequently, very few birds. The centrepiece of the park was a railway carriage which broadcast mournful music in memory of those who had died in WWII. I was not mentally alert enough to get maximum benefit from this walk back into history.

My two over-riding impressions of Japan were first the cleanliness, which I have already mentioned, and secondly the friendliness of the people. Since I spent more time in Yokohama, I will give an example of my experience there.

Narita airport is 60 km, 37 miles, east of Tokyo city centre. The quickest way into the city is by the Narita Express which proceeds first to Tokyo and then on to Yokohama. There is also a coach service which drops passengers off at the main hotels. However, the service is not as frequent as the trains. There are, of course, taxis but taxis are for people with more money than sense! I could have taken a coach directly to my hotel in Yokohama but would have had to wait for some four hours before the next one was due. There was a train leaving in 20 minutes so I took it.

I stepped off the Narita Express at Yokohama station, one of the biggest stations I have ever been in, and stood trying to find my bearings. With four possible exits (north, south, east and west) it was difficult to know which one to use. I assumed it didn't matter too much since what I really wanted was a taxi to get me to my hotel. As I looked around for a taxi sign, I was approached by a man who asked me, in good English, where I was coming from and where I wanted to get to. He wasn't sure of the location of my hotel so he escorted me to the tourist information point, made enquiries on my behalf, and then escorted me to the appropriate exit and hailed a taxi. He shook my hand, welcoming me to Japan and hoping I would enjoy my stay, before hurrying off about his own business. This was typical of all of the people I met and spoke to. They were always helpful and courteous. The language barrier meant that they did not always fully understand what I required but they always went out of their way to be helpful.

While there were large crowds everywhere I went, I must say also that I never felt under any kind of threat in Japan. Despite following the global trend of rising crime rates, Japan was still one of the safest countries in the world. I took the normal precautions against pickpockets and bag snatchers but didn't feel the need to be super-vigilant as I would in some of the other cities I have visited.

One thing about Japan, which I had difficulty adapting to, was the incessant bowing. For me, and I expect for most Westerners, bowing has strong cultural associations with servility and inferiority. I know that is not the case in Japan where bows are exchanged to convey mutual respect. When someone bows to you, you should really bow back. "When in Rome…" seems an easy enough principle to adopt but I found it hard to bow in Yokohama. Bowing made me feel uncomfortable. I would have preferred a handshake. But bowing came into its own in 2020 following the outbreak of Covid-19 when we were advised that handshakes were much more likely to spread the virus.

Apart from the things I have mentioned, I didn't see much of the real culture of Japan. I have no taste for sake, or for sushi, and had neither the time nor the inclination to visit nightclubs.

There were approximately 3,334 delegates from 138 countries attending our Congress. The four-day Congress was sponsored by UNICEF, the Japanese Government, and ECPAT. Attendance was by invitation only. Apart from Government delegates there were representatives from 21 international institutions, including UNICEF, and 148 NGOs from all around the world, including the IAYFJM which I was representing.

The main objectives of the Yokohama Conference were to review progress of specific action plans assigned to all the participating nations at the first conference in Stockholm in 1996, and to call for additional action if sufficient progress had not been made towards these goals.

I report on both the Stockholm and the Yokohama Conferences in the final chapter – A World Apart – so I will not comment further on them here.

Jordan
2007

In 2007, the United Nations Office on Drugs and Crime (UNODC) appointed me as *External International Expert* (Team Leader) for a *Mission* to Jordan. The task was: to evaluate the Juvenile Justice Reform Programme entitled – *"Strengthening the legislative and institutional capacity of the juvenile justice system in Jordan"*. This entailed document review, field visits and interviews with key interlocutors and key players in Jordan. The report was to be completed by the end of June 2008.

I found a lot of friction between the Jordanians and the UN representatives. It appeared to me that the root cause was a change in the traditional approach to funding projects. Traditionally a country would submit a project for consideration. If approved the UN would transfer funds to the country concerned and then monitor how it was spent. In the case of Jordan, the department concerned with the project – the UNODC – decided not to transfer the funds but rather to pay for each part of the project as it was completed. This didn't go down well with the Jordanians. They saw it as an insult, a lack of trust in their integrity.

The UNODC needed to appoint a diplomat to oversee the project. They didn't. The first meeting was a heated affair with accusations being thrown backwards and forwards. The UN rep recorded every word in the minutest detail so that when the minutes were read out at the next meeting the war broke out again. And so, it went on.

Much useful work had been done and a number of objectives achieved. Much important work remained to be done but the possibility of satisfactory completion seemed remote because relationships had been soured. In my view both sides had to share the blame but the heavier burden had to be shouldered by UNODC as the senior partner.

My report didn't go down well. UNODC was unhappy and put a lot of pressure on me to change it. I said I would be happy to change things if they could identify errors of fact. However, it appeared that what they wanted me to do was to rewrite the report piling all the blame on the Jordanians. The UNODC country rep in Jordan resigned before I completed my report and took up a job in the Sudan. This appeared to me to be an admission of responsibility.

I had been looking forward to my visit to Jordan. There was so much I wanted to see. *Bethany Beyond the Jordan*, where John the Baptist preached and where Jesus was baptised is located on the Jordan side of the river. *Mount Nebo*, where Moses viewed the Land of Canaan is located in western Jordan. In northern Jordan there is a small creek where an angel met and wrestled with the patriarch Jacob. The rock struck by Moses to bring forth water, and the patriarch Aaron's tomb, are both in southern Jordan. Madaba, a city south of Amman, is the site of a large ancient church with detailed mosaic tile work. There are many Arab and Frankish Castles from the period of the Crusades. And, of course, there is the Dead Sea. I was based in Amman so I expected to see at least some of these. However, my schedule was so tight that I only had time to work, eat and sleep. I was told that if I needed time to consider how best to rewrite my report, my flight home could be delayed to allow some time for sightseeing. I was not about to compromise my integrity, however much I wanted to see the holy sites. I was not invited to do any more work for UNODC.

Kosovo

2000

The UN took control of Kosovo in the year 2000 when the war with Serbia was finally brought to an end. Judges who had been sacked by the Serbs some twelve years previously were reappointed and had to undergo a period of refresher training. I was called in as a Council of Europe Expert on the European Convention on Human Rights (ECHR). All Council of Europe member states are party to the Convention and new members are expected to ratify the convention at the earliest opportunity. The UN had ruled that Kosovo would be regarded as a new member state. I was asked to conduct three one-week seminars during March and April to bring the judges up to speed on the ECHR. This was one of the toughest assignments – *Missions* in UN parlance – I was ever asked to undertake.

I was based in Pristina, the capital, and had the first seminar there. The city had suffered greatly during the war and many of its grand buildings were in ruins. The airport was still out of action at the time of my first visit. All the streets and main roads were full of potholes – some big enough to swallow a small car, which is probably why everyone drove massive four-wheel drive jeeps. It hadn't rained for some time and everything, including my shoes when I went for a walk, was covered in white dust. The electricity was functioning again intermittently but the few stores and restaurants still open had their own generators. These kicked in each time there was a power cut – which was quite often. The inhabitants appeared to be vastly outnumbered by the various UN agencies and a wide range of Non-Government (i.e., voluntary) Organisations (NGOs). Someone told me that there were almost 100 UN and voluntary aid agencies. The place was eerily quiet, apart from the throb of the many diesel engines. At first, I wondered why. Then I suddenly realised that there was no bird song. The birds had apparently deserted the place and hadn't, as yet, returned.

My hotel was spartan but quite comfortable. The city was surprisingly peaceful, although I was told to be careful as Serb sympathisers were likely to launch attacks if the opportunity arose. I found an excellent little cafe where I liked to sit and relax in the evenings after work. Because the power cuts were so unpredictable, we mostly dined by candlelight which enhanced the atmosphere. I quite liked Pristina and thought it must have been a nice city before the war.

My second seminar was in Pec, a city about one hour's drive to the south. It was very hot as we drove south and, as there was no air conditioning in our minibus, we stopped for a break. The Serbs had launched a scorched earth policy and had destroyed everything in their path – torching houses so that the Kosovars could not return to them. The house we stopped beside was derelict. I went round the back to view the architecture at the rear of the building, as men are wont to do on long bus runs. I got a severe telling off. I was informed that the Serbs frequently left landmines round the back of buildings to kill the unwary.

The third seminar was in Mitrovica some 25 miles to the north. Mitrovica is divided by the River Ibar – 80% Serb to the north and 20% Kosovar to the south. The venue for my seminar was located close to the main bridge over the river and, during a lunch break, I slipped over the bridge to get a photograph from the other side. Again, I got told off. I pointed out that the war was over but was advised that the signing of a document did not guarantee that hotheads, on either side, would not take a pot shot at someone they believed to be the enemy. 50,000 recalcitrant Serbs in northern Kosovo said they would never accept Pristina's authority. For some the war continues.

I said above that these missions were the most difficult I ever undertook, and it was not because I got told off on two occasions! The judges were difficult to work with. I was accustomed to working with an interpreter. This does not normally prevent meaningful interaction. The judges here were not interested in interaction. They were resentful at being made to do a refresher course since they had all been experienced judges before being sacked by the Serbs. I was there to teach them about the ECHR. They didn't want to know.

I tried to get them to focus on human rights by telling them that, in their role as judge, they would meet a cross-section of the community – some Kosovars, some Albanians, some Serbs. All should be treated equally, irrespective of religion or ethnicity. One judge told me that Serbs appearing in court should always be found guilty. Their property should be confiscated and handed over to a member of the Kosovar community who had had his/her property destroyed by the Serbs during the war. This seemed to be the majority view. I didn't really blame the judges for their lack of interest in the ECHR. In my view the course came too soon for them. Their emotions were still very raw. The UN would have been better advised allowing them to settle back into their role as judges for a period before introducing them to the European Convention on Human Rights.

Kosovo remained under UN control – an internationally supervised limbo – for nine years during which time I expected to be back. I was advised that my trip was cancelled on receipt of information of a planned attack on the international judges in the Appeal Court. On February 17, 2008, Kosovo, with a population of under two million, declared independence from Serbia. It is difficult to see the benefits. Kosovo's independence has been recognised by just 111 of the UN's 193 member states. Crucially, Russia, China, Serbia and five EU countries do not recognise it. In February 2018, the Prime Minister was refused a visa to visit Britain. He had previously been refused a visa for the United States. It appears that, despite independence, it is almost impossible for Kosovars to travel.

Liberia
2010, 2011

I took on a major project in Liberia, West Africa, in 2010. Liberia was originally an American colony. In 1816 a group of white Americans founded the American Colonization Society to deal with the "problem" of the growing number of freed slaves by resettling them in West Africa (the majority of slaves were of African origin). Liberia was established as a settlement in 1821 on land purchased from indigenous leaders. The capital Monrovia, where I would be based, was founded in 1822. It was named in honour of James Monroe, then President of the United States, and a prominent supporter of the colony. Liberia's current population of around four million is made up mainly of indigenous Africans. Only a small percentage are descendants of former American slaves.

Liberia was a peaceful country until 1980 when a military coup led by Samuel Doe ousted the Tolbert government, with many people being executed. Violence erupted again when a civil war broke out in 1989 and lasted until 1996. The city was severely damaged, notably during the siege of Monrovia, with many buildings and nearly all the infrastructure destroyed. Major battles occurred between Samuel Doe's government and Prince Johnson's forces in 1990 and with the *National Patriotic Front of Liberia (NPFL)'s* assault on the city in 1992. About 200,000 people were killed during that first civil war.

A second civil war led by Charles Taylor broke out in 1999 and lasted until 2003. The city suffered further damage. Samuel Doe and many government ministers were executed. Charles Taylor appointed himself as President. He would later be cited for war crimes and crimes against humanity. On 30 May 2012 he was sentenced to 50 years in prison for war crimes. Readers may remember the trial more for the story that Charles Taylor gave Naomi Campbell a number of uncut diamonds after a 1997 dinner in South Africa – a present which Naomi dismissed as "just a handful of dirty stones"!

A legacy of the two civil wars is a large population of homeless children and youths, either having been involved in the fighting or denied an education by it. There are also a large number of young men with missing limbs, mostly deliberately chopped off as acts of war.

This was the city where I arrived at the beginning of October 2010. The destruction wrought by the wars was visible everywhere. I remember thinking that if I were an architect, I wouldn't have known what could be salvaged. The easy option would be to level the buildings and start from scratch. I would later find out that many people had no running water and no electricity. Where there was electricity, it tended to be a bit intermittent – it was not unusual for all computers to be down, maybe for the whole day. I didn't see any traffic lights working throughout the entire city. The roads were in a mess with large potholes everywhere. There was one huge pothole close to the UNICEF office. Someone had stuck a branch of a tree in it so that motorists would know to keep well clear. The hole had not been repaired when I returned four months later!

Monrovia has a tropical monsoon climate. The climate features a wet season, May to October, and a dry season, but sometimes it rains even during the dry season. We tend to think that Belfast gets a lot of rain. On average it rains in Belfast on 213 days per year but the total average annual rainfall amounts to 33.3 inches. Monrovia averages 208.7 inches of rain per year! I arrived just at the end of the wet season. Luckily most of the rain fell at night, commencing with violent thunderstorms and torrential downpours. I was amazed to see people out in the morning washing clothes in the potholes in the road which were now filled to overflowing with water. Not only did the people not have running water, they did not even have containers to catch water when it rained. Their clothes were then spread along the sides of the road to dry in the hot sun. The average temperature throughout the year is 26°C, but, during my visit, it remained in the low 30°s C.

Many people lived in ramshackled "houses" along the sides of the roads. The "houses" consisted of sheets of rusty corrugated iron for the roofs and the walls with no proper doors or windows. If farmers kept animals in shelters like that at home they would probably be prosecuted! I was amazed at how resilient the children were as I watched them playing happily in the sun.

There is quite a nice beach in Mamba Point but I found I was unable to walk on it. It was the home of a large number of people who slept rough there. During the day, they had their clothes spread all over the beach drying in the sun.

My hotel was in a beautiful setting overlooking the bay. Most mornings as I was having breakfast a tanker drew up on the road below to deliver drinking water to the hotel. The tanker was old and the hose was full of holes. I watched children running with whatever containers they could find to catch the water escaping from the holes and take it home for drinking. I was amazed at how orderly they all were, queuing up and waiting their turn.

The first thing I had to do when I first arrived at the UNICEF office was to report to the Security Officer where I got an hour-long lecture on what I should and shouldn't do and where I should and shouldn't go in Monrovia. I was told that the city of Monrovia consists of several districts, most of which were unsafe. The only safe area was Mamba Point, to the west/southwest of downtown. This is the city's principal diplomatic quarter, and home to the Embassies of the United States and United Kingdom as well as the European Union Delegation. The UNICEF offices are here, as is my hotel. Mamba Point was the only area where I could wander around fairly safely. If I wanted to go out on the town for a drink or a meal in the evenings, I should advise UNICEF in advance and they would provide transport. Local taxis were unsafe.

South of the city centre is Capitol Hill, where the major institutions of national government, including the Temple of Justice and the Executive Mansion, are located. While I was based in the UNICEF office, I would be doing most of my research in the Temple of Justice (which housed the courts and the Judicial Training Institute). At the southeasterly base of the peninsula is the independent township of Congo Town, and to its east is the large suburb of Paynesville. I would be taken there for a day's observation in one of the Magistrates' Courts.

Finally, I would be issued with a pass. I should wear it at all times. I would not gain entry to the UNICEF offices, or the Temple of Justice, or any other Government building without it.

That was the most intensive security briefing I have ever received. But the practice did not quite match up to the theory. I didn't receive my security pass until a few days before departing for home – and that was only after intensive pressure from me! I never had any problems gaining access to any of the buildings. I wondered would it have been so easy had I been black skinned. I didn't test the pledge that UNICEF would provide transport in the evenings if I wanted to go out because I don't frequent pubs and don't like eating late in the evening. I spent my evenings working in my hotel room. But I was a bit sceptical when I learned from one of my drivers that they stopped work at 5 pm!

Each morning I was picked up at my hotel and taken across town to the Temple of Justice. I would ring my driver when I was ready to come back and he would come and collect me – providing it was not after 5 o'clock. I got a bit of a shock on Friday of the first week when I got no response to my phone call. I had no idea how to get back to my hotel. I spoke to one of the judges. He was surprised that no one had told me that the drivers got a half day each Friday. He arranged for someone to drop me off.

I found the people in Monrovia very friendly and helpful but police and military very edgy. I never wandered very far but I am a keen birdwatcher, and I wanted to get out a little. I went for a walk close to the Temple of Justice. I took a photo of the building itself and then decided to take some pictures of the Parliament building and of the Executive Mansion. I realised that I was being closely observed by some soldiers in a jeep. Then I spotted a notice which said "no photographs" so I headed back.

The highest security was reserved for the American Embassy which I walked past several times each day. Apart from the very high walls and security guards on the gates the road in front of the building had permanent roadblocks. Traffic had to take an alternative route or endure intensive security checks. There were also large notices on the walls saying, "Strictly No Photographs".

I had got some excellent photographs of birds from the hotel balcony – particularly of two kingfishers which seemed to have a nest nearby. I also had some nice photos of lizards. But I had seen some beautiful male lizards close to the UNICEF offices and I wanted some photos of them. I was down on my knees photographing a lizard when I noticed a police officer coming towards me. As he got closer the lizard ran off. I looked up, laughed, and said: "you frightened it away on me". The policeman wasn't smiling. He said: "I'm afraid you will have to come with me". I asked him why and he replied that it was not permitted to take photographs. I asked him why and he replied: "There is a notice on the wall". We weren't near any walls so I asked him if he meant the notice on the wall of the American Embassy which was several hundred yards away. He did. I pointed out that I was not photographing any buildings. He could see as he walked up to me that I was photographing a lizard. I showed him all the photographs in my camera to confirm what I was saying. He still insisted that I go with him. I walked with him up to the gate of the American Embassy. He rang for his superior to come out, rank of sergeant – I guessed. I went over the whole discussion again. She said she couldn't make a decision and called for the inspector. He agreed that all the photographs I had taken were completely innocuous but said he would still need a statement from me. I provided him with my name, address, date of birth, country of origin, what I was doing in Liberia, where I was staying, who I was working for. I was able to show him my UN pass, which, fortunately, I had received by now. Eventually the inspector apologised for detaining me, said he was sorry for any inconvenience he had caused and thanked me for being so understanding. I was free to go.

The project I had come to work on was expected to take three to four months running from September through to December. In the event, it was not completed until February 2011.

My task was to develop a child-justice curriculum and training manual for magistrates, judges and prosecutors that would include child rights, the international standards on juvenile justice, forensic interviewing of children as victim, witness and/or offender. I was to conclude the project by delivering a *Training of Trainers* workshop with a view to teaching a small group of trainers how to use the manual. They would then be responsible for delivering the training to their colleagues around the country by way of *cascade* training.

I was back in Monrovia at the beginning of February 2011 for the *Training of Trainers* seminar. All went well and everyone was pleased with the work I had done. I kept in touch with colleagues in Liberia and they informed me that the cascade training was going well also.

Mexico
1981, 2006

Tijuana

I first visited Mexico in 1981. My son Liam and I were in LA and set aside one day to take a bus trip down the coast to San Diego. San Diego shares a 24 km border with its sister city Tijuana, which has been nicknamed "The Gateway to Mexico". It is the most visited border city in the world. Current estimates are that more than fifty million people cross the border between these two cities every year. We didn't know when we might get another opportunity to visit Mexico. We couldn't pass this chance up.

Tijuana is very much a touristy border town, with souvenir shops everywhere. We didn't get beyond the bustling main street, Avenida Revolución. We spent our time looking around the shops. We bought a few souvenirs. Everyone was haggling over prices. I managed to bargain the price of a nice leather wallet for my brother-in-law all the way down from $45 to $22! All prices were in dollars – Liam had to specifically ask for some Mexican coins. We took a few photographs. A man with a horse and cart wanted to charge me a dollar for taking a photo! I had a coffee and Liam had a fruit juice. Then it was time to cross over the border to San Diego again and catch our bus back to LA.

Mexico's Ancient City, Teotihuacán

I was back in Mexico in 2006. I landed in Mexico City, in April that year, on my way to a conference in Guadalajara. I had a day and a half free before going on to the conference. I wanted to make the most of it. Top of my list of things to see was Mexico's ancient city, Teotihuacán, 35 miles (56km) north of Mexico's present-day capital. My interest focused on the fact that it is the site of many of the most architecturally significant Mesoamerican pyramids built in the pre-

Columbian Americas. I was too late to join an official tour so I caught a bus out to the site, bought a guidebook and set off to explore it on my own.

Construction at Teotihuacán began around 150 BC, and continued until 250 AD. At its height, the city covered 21 square miles and was home to as many as 200,000 people, accommodated in multi-floor apartment compounds. It was the largest city in the western hemisphere. Its major monuments were sacked and systematically burned around 550 AD. It is not known who was responsible. It was abandoned some time thereafter.

The largest building in Teotihuacán is the Pyramid of the Sun. It is believed to have been constructed about 200 AD. It is 224.942 metres (738 feet) across and 75 metres (246 feet) high. The ancient Teotihuacanos finished their pyramid with lime plaster imported from surrounding areas, on which they painted brilliantly coloured murals. While the pyramid has endured for centuries, the paint and plaster have not and are no longer visible.

The pyramid is part of a large complex in the heart of the city. The city's broad central avenue – the Avenue of the Dead – runs down the middle of the site, linking the Pyramid of the Sun with the Pyramid of the Moon and the Temple of Quetzalcoatl. The avenue is lined with many smaller talud-tablero platforms. The Aztecs believed they were tombs, inspiring the name of the avenue. It is now believed that these were ceremonial platforms, once topped with temples.

I spent a pleasant couple of hours climbing to the summits of the Pyramid of the Sun and the Pyramid of the Moon to have panoramic views of the city and strolling along the Avenue of the Dead trying to imagine what life was like for the people who lived there. It was then time to catch a bus back to Mexico City.

Next morning, I decided to have a stroll around the streets close to the hotel. I found the main road closed to traffic. With the media always focusing on negative stories about Mexico City, I wondered if there had been a traffic accident or a shooting. Then I heard music and a band appeared in the distance heading a parade. There were, in effect, a number of bands, interspersed with groups of people carrying banners. I discovered that it was the feast day of the local patron saint and these good people were on their way to the local church to celebrate mass. It was not the kind of story the media were likely to be interested in. I returned to my hotel for lunch and then headed for the airport to catch my flight to Guadalajara.

The entry in my guidebook said: "Far from the well-trodden tourist trail, the colonial city of Guadalajara is one of Mexico's most overlooked destinations. Situated in the western state of Jalisco, Guadalajara is Mexico's second-largest city, and the birthplace of two of its most emblematic exports: tequila and mariachi music. It is sunnier and less overwhelming than Mexico City, while offering better value for money and a more "Mexican" experience than better-known tourist resorts." Unfortunately for me, I had used up all my spare time. There was no time for sightseeing. I was here to work.

Myanmar or Burma?

2001, 2002, 2003

Why Has This Country Got Two Names?

Burma declared independence from Britain on January 4, 1948, ending 60 years of colonial rule. When the military seized power from General Ne Win's government in 1989, there was a push to establish a national identity among the country's assortment of ethnic groups. Burma is considered to describe ethnic Burmans only, so Myanmar became the politically correct term, which is supposed to encompass all who live in the country. Many names across the country changed. For example, the city of Rangoon became Yangon, Moulmein became Mawlamyine, Ayeyarwady became Irrawaddy. The name changes were also a way to rid the country of British colonial influences. Officially, the correct name is "Myanmar" but the country's democracy movement prefers the name "Burma" because they do not accept the legitimacy of the unelected military regime to change the official name of the country. Internationally, both names are recognised, and they are frequently used interchangeably.

In 2001 I was in Myanmar, part of a team of four invited to deliver a judicial training programme over a three-year period – 2001 to 2003. Each of us would focus on our own area of expertise. I would focus on the Convention on the Rights of the Child (CRC) and associated international instruments.

We had four training seminars in total. The first two, November 2001 and July 2002, were held in the capital, Yangon (Rangoon). The third seminar was held in Mawlamyine, about 100 miles east of Yangon (close to the border with Thailand) on 20 and 21 February 2003. Mawlamyine is the capital of Mon State and has a population of almost 300,000 people. The town's signature landmark is *Kyaikthanlan Pagoda* built in 875 AD and thought to be the site from where Rudyard Kipling wrote his famous poem, '*The Road to Mandalay*'. The fourth seminar was in Myitkyina in north-eastern Myanmar on 25 and 26 February

2003. Myitkyina lies on the Irrawaddy River, 920 miles north of Yangon and 488 miles north of Mandalay. It is the capital city of Kachin State.

Yangon, with a population of about four million people in 2001, was Myanmar's capital[25]. I couldn't describe it as a beautiful city. It had a rundown appearance with many buildings – apart from the pagodas – in need of renovations and all needing a lick of paint. The effect of the sanctions was clearly visible here. I was keen to see as much of the city as time allowed.

I visited the magnificent Shwedagon Pagoda which is believed to be the oldest in the world, having been built, according to legend, about 2,600 years ago. It is certainly the biggest and the most impressive. Rising from the top of a hillock, its golden spire soars to a height of nearly 100 metres to dominate everything around it. It is also the most sacred Burmese pagoda, housing relics from all four Buddhas. It is covered in 60 tons of gold and is 325 feet tall. The crown (or umbrella) is said to have over 5,000 diamonds and over 2,000 rubies, with a 76-carat diamond bud at the very top. With a 14-acre terrace the Pagoda is like a small village. I had to remove my shoes and socks before entering. The Pagoda is surrounded by prayer halls which were filled by many local people who had come here to worship.

There are planetary stations representing different days of the week. Each station is placed in a specific location such as north and has a Buddha image, a guardian angel, a planet and an animal representing that day. Visitors are encouraged to present an offering and light a candle to the Buddha for the day of the week on which they were born, pouring water over the Buddha image, the guardian angel and the animal, with a prayer or a wish. Having been born on a Wednesday, the animal for my day of birth is an elephant, although things are a little bit complicated because I don't know whether I was born in the morning or the afternoon. The animal for Wednesday morning is an elephant but for Wednesday afternoon is a *tuskless* elephant. I was not aware that there are tuskless elephants but, apparently, they are fairly common, particularly in Sri Lanka.

---

[25] The military government relocated the administrative functions to the purpose-built city of Naypyidaw in central Myanmar in 2006.

The Road to Mandalay

At the end of our first mission to Myanmar (November 2001) my colleagues and I decided to go on a short tour. We were told by the authorities that, as honoured guests, we were free to go where we liked. No restrictions would be imposed.

Perhaps the most pleasurable way to see Myanmar, feel its pulse, live its legends and understand its history, is to cruise the Irrawaddy River. The river has long served as an economic and spiritual lifeline for the country. While enjoying the river's tranquillity, life on the riverbank offers endless fascination and the possibility of gaining a deeper appreciation of Myanmar's culture, history and people.

Flowing from north to south through Myanmar, the Irrawaddy River is 1,300 miles long. So, however appealing, it was not possible to cruise the full length of the river. Since it is often referred to as the "The Road to Mandalay", after Kipling's famous poem, we decided that Mandalay should be our goal.

I chose the Irrawaddy River Delta as the topic for a school geography project many years ago. I never imagined that I would take a cruise on the river one day. It was like a dream world as I stood on the deck looking at the small teak and bamboo dwellings along the riverbank, excited children waving to the passengers, the women doing their washing in the river, the oxen cultivating the fields.

The Irrawaddy River is Myanmar's lifeline. I stayed long into the evening watching the endless flow of ferries, bamboo rafts, barges and fishing boats, all plying their trade along its waters. I was back early in the morning to watch the sunrise while most passengers were still asleep. I watched elegant monasteries and ruined ancient temples rise above canopied trees still shrouded in mist while the sun glistened on the golden domes of the many pagodas. I was saddened at the sight of endless piles of teak logs plundered from the hardwood forests and ready to be loaded on barges to begin their journey to China. Was it any wonder the Government was not feeling the effects of the sanctions?

The last leg of our journey was by minibus. Before heading for Mandalay, we were taken on a short tour of Bagan the ancient capital of the Kingdom of Pagan. During the kingdom's height, between the 11th and 13th centuries, over 10,000 Buddhist temples, pagodas and monasteries were constructed in the Bagan plains alone. The remains of over 2,200 temples and pagodas survive to the present day.

Pagodas and Stupas are effectively tombs. They were originally built to preserve the remains of Sakyamuni, the founder of Buddhism. All contain Buddhist relics. Some people claim that there is no difference between a pagoda and a stupa. My understanding is that a stupa is a solid building with no entry points. A pagoda has one or more entry points to allow access to prayer rooms. Pagodas may also be built in grottoes or temples for offering sacrifices to your ancestors. Bagan is an amazing sight with pagodas and stupas stretching as far as the eye can see.

It is only a short distance from Bagan to Mandalay. Mandalay is considered to be the centre of Burmese culture and Buddhist learning. It was the country's last royal capital. With a population of about one million it is the country's second largest city. It has many historic points of interest but somehow it didn't strike a chord with me. It was not the city of Mandalay which had attracted me but rather the journey there.

I did want to visit U Bein Bridge in Amarapura. Amarapura, a former capital of Myanmar, is now a township of Mandalay. The U Bein Bridge is a 1.2 kilometre wooden footbridge, the longest teak bridge in the world. It was built by the Mayor of U Bein. He salvaged unwanted teak columns from the old palace when the capital was being moved to Mandalay. The bridge still serves as the most important communication link for the people of his villages. It is still used by monks, fishermen and villagers going to and from work. Some villagers have set up stalls in the rest points that dot the crossing. Most tourists walk across the bridge and back or take a boat ride along Taungthaman Lake.

I sat on the riverbank, beside the lake, surrounded by children. I was intrigued by their signs, which I didn't understand at first. They would run their fingers through their hair, or rub their teeth with their forefinger, or make a sign as if they were writing. Then I realised that they were asking for combs, or toothbrushes, or pens. The best hotels always provide these things in case guests have forgotten to bring their own and tourists who have been here before take some to hand out to the children. I hadn't even thought of it. All of the children were selling trinkets. Some of the boys had taken young birds, which hadn't learned to fly yet, out of their nest and were offering to "sell" them to the tourists. The tourists could then set them free. I have no doubt the children would promptly catch them again to sell to the next tourist.

I was aware that English is taught in the schools in Myanmar but was still surprised at how well these children could converse. The teacher in me took over and I sat talking to them about the importance of going to school – clearly, they were missing school when they were here at the bridge selling trinkets. I advised the boys that they shouldn't take young birds out of their nest, as they wouldn't survive without their parents. I knew I wasn't getting through to the children. I bought a few trinkets and released a few birds so everyone was happy.

Getting to Know the People

Most of our contact in Myanmar was with senior army personnel, senior police officers and senior judges. When we were invited out for meals the restaurants we went to were as good as the best in Paris, Rome, London or New York. There was no sign of sanctions hurting at this level. We had very little contact with ordinary people. It seemed clear that that was how the Junta wanted it to be.

Our trip to Mandalay had given us an insight into how tightly the Junta controls things. When we requested permission to cruise up the Irrawaddy, we were told that, as honoured guest, we had freedom to go where we liked. There would be no restrictions on our travel. We had just checked into our hotel in Mandalay when we received a phone call from the Minister of Justice inviting us to have dinner with him on our return. Clearly, despite being guests of honour, free to go and come as we pleased, someone had been keeping close tabs on us. They knew exactly where we were. We flew back to Yangon next day and duly met the Minister for dinner.

Buddhism

I knew nothing of Myanmar or its peoples before my series of visits. It didn't feature in my studies of history apart from occasional mention of the war in Burma. I had a gentle image of Buddhism defined by the self-effacing words of the Dalai Lama, the global popularity of Buddhist-inspired meditation and postcard pictures of crimson-robed, barefoot monks receiving alms from villagers at dawn.

This image was reinforced for me during my four missions. I learned that the village monastery functions as the centre of social activities, where most communal affairs take place. Where necessary, monks take the initiative in various social projects, mainly by giving guidance and leadership, such as the construction of schools and hospitals, roads and small reservoirs, and at times even the digging of a village well. They also take the lead in raising funds for

147

such projects. In times of natural disaster monks often provide the most effective leadership in pooling resources together to help ameliorate the suffering of victims and their families. The monks are often instrumental in setting up charitable programmes or foundations for the welfare of society.

School facilities are often located on monastic property with monks offering their teaching skills free of charge. Monastery grounds are turned into playgrounds for the children.

The monks take a keen interest in protecting the environment, focusing, in particular, on helping to preserve the forests and water resources of the country. In most cases their influence and intervention in such matters prove more effective than those of government agencies.

My role during the four missions was to provide judicial training. I was keen to learn what kind of offences the judges dealt with and whether minor offences were diverted from the court system. I was informed that poor villagers often turn to monks to mediate land disputes, family problems, and differences among neighbours. Appeals are heard by the village chief or elders. Only the more serious offences end up in court. It was unusual for young offenders to end up in prison. Unruly and difficult children are taken to a monastery for training in discipline and other social values.

By the time my series of training programmes had come to an end I had a very positive view of Buddhist Monks. Later something happened which dented their clean-cut image. I was on my way by train from Bangkok to Kanchanaburi. The carriage I was in was practically empty – just myself and a Buddhist Monk. He took out a takeaway lunch, which, presumably he had bought in Bangkok. I was surprised because I knew that the monks generally eat only once a day. However, I also understood that the practice is a voluntary one, so he wasn't breaking any rule. Besides he appeared so ravenous that no one could have condemned him. But I did condemn him for his next actions. As he finished each part of his meal, he threw the rubbish out of the window. So much for concern for the environment! My admiration for this Buddhist Monk took a nosedive but I had been around long enough to know that there are delinquents in all organisations.

Hatred of Muslims

I thought I knew a lot about Buddhist monks. In fact, I knew very little. I was totally unaware of a monk called Ashin Wirathu who presided over a very large

148

monastery of some 2,500 monks. He rose to prominence in 2001 – the year I arrived in Myanmar – when he created a nationalist campaign to boycott Muslim businesses. He was jailed for 25 years in 2003 for inciting anti-Muslim hatred but freed in 2010 under a general amnesty. Since his release, he has gone back to preaching hate. He calls himself the "*Burmese Bin Laden*". He has thousands of followers on Facebook and his YouTube videos have been watched tens of thousands of times. In that year alone Buddhist lynch mobs killed more than 200 Muslims and forced more than 150,000 people from their homes.

The hatred of Muslims has a long history, dating back centuries, in Myanmar. Coincidentally, on September 22, 1938, just three weeks after I was born, the British Government set up an enquiry into the assault and massacre of Muslims and the destruction of their property.

So much for the gentle image of Buddhism! The myth is now shattered. For me Ashin Wirathu revealed both the gentle image and the darker side of Buddhism in a comment to reporters following a two-hour sermon laced with hate.

"*You can be full of kindness and love, but you cannot sleep next to a mad dog,*" he said, referring to Muslims.

Tension Between Buddhists and Rohingya Muslims

Myanmar, population 54 million, is predominantly Buddhist. 1.33 million Rohingya Muslims are concentrated in the western state of Rakhine, one of the remotest, poorest, and most densely populated areas of the country.

Tensions between the Buddhist majority and the Rohingya date back to the beginning of British rule in 1824. As part of their divide-and-rule policy, British colonists favoured Muslims at the expense of other groups, causing deep resentment amongst Buddhists. A military coup in 1962 ushered in a new era of repression and brutality against ethnic minorities like the Rohingya who were effectively rendered stateless in 1982.

The Myanmar government regards the Rohingya as illegal immigrants from the Indian subcontinent and refuses to give them citizenship status, effectively making them stateless. They cannot travel, get married or seek medical treatment without official permission. Rights groups say the Rohingya also face persecution, forced labour, land confiscation and limited access to education.

Aung San Suu Kyi

Aung San Suu Kyi is the daughter of independence hero, General Aung San, who was assassinated in 1947, and the rightful leader of Myanmar. She led the opposition to the military dictatorship and her political party, the National League for Democracy (the NLD), won a landslide victory in a general election in 1990. The Generals ignored the result. They refused to hand over power. Suu Kyi had been under house arrest since 1989 and remained so for 15 of the next 21 years. Many of her supporters were killed, thousands were imprisoned. She was awarded the Nobel Peace Prize in 1991 for her non-violent struggle for democracy and human rights but was unable to go to Oslo.

The Generals were under increasing international pressure to restore democracy and accept the election results in return for the easing of sanctions. Aung San Suu Kyi was unexpectedly released from house arrest on May 6, 2002. When my colleagues and I arrived in Myanmar for our second mission in July that year, we were offered the opportunity to have dinner with her. We were not about to turn down the offer to meet the esteemed Nobel Laureate. I wanted to ask her for her views on our judicial training programme in her country.

I found Suu Kyi to be such a charming lady, totally unassuming, and not the least bit bitter about her treatment, or that of her supporters, at the hands of the Junta. She is totally committed to a non-violent struggle for democracy and human rights. When asked how it felt to be free after a period of house arrest, she replied: "*my mind is always free*".

Suu Kyi said she supported our initiatives in Myanmar. Anything we could do to promote the best interests of children was welcome. The judges needed to be trained in the CRC. The development of new legislation which would incorporate the CRC was vital (one of my colleagues was working in this area). She warned us not to be too disappointed if things did not work out exactly as planned.

I found Suu Kyi to be one of the most remarkable ladies I have ever met. It was an honour to have dinner with her and to have a long informal chat. When we were leaving, she put her arms around me and gave me a hug and a kiss. She said the Irish have always been very supportive of her efforts.

We left in high spirits, hopeful that her years of sacrifice were beginning to bear fruit. Little did we know that the storm clouds were gathering once more and the age-old tensions between the Buddhist majority and the Rohingya Muslims were threatening to undermine all her good work.

The Generals called a general election in 2015. The NLD won another landslide victory, taking 86% of the seats in the Assembly of the Union. But the Military Junta had no intention of giving up power. Suu Kyi was prohibited from becoming President. Instead she was given the newly created role of State Counsellor, a role akin to a Prime Minister or a head of government. The Junta retained overall control.

The conflict between Buddhists and Muslims was becoming increasingly hostile and in August 2017 Myanmar's military launched a violent crackdown on the Rohingya. The violence included the killing of thousands of people, the rape of women and children, and the razing of villages. More than 700,000 Rohingya fled to Bangladesh. The UN described it as ethnic cleansing and possible genocide.

The government denies widespread wrongdoing and says the military campaign across hundreds of villages in northern Rakhine was in response to attacks by Rohingya insurgents.

Since ascending to the office of State Counsellor, Aung San Suu Kyi has drawn criticism from several countries, organisations and figures over her alleged inaction to the persecution of the Rohingya people in Rakhine State and refusal to accept that Myanmar's military has committed atrocities as part of a programme of ethnic cleansing. There have been calls for her to be stripped of her Nobel Prize. Commentators who called her *The Nelson Mandela of the East* are now quick to contrast his achievements with her lack of achievement. They fail to consider the different contexts.

When Frederik Willem de Klerk succeeded P W Botha as President of South Africa in 1989, he was aware that growing ethnic animosity and violence was leading the country into a racial civil war. He decided to end the system of racial segregation. He permitted anti-apartheid marches to take place, legalised a range of previously banned anti-apartheid political parties, and freed imprisoned anti-apartheid activists, including Nelson Mandela. He negotiated with Mandela to fully dismantle apartheid and establish a transition to universal suffrage. In 1993, he publicly apologised for apartheid's harmful effects. He oversaw the 1994 multiracial election in which Mandela led the ANC to victory. His own National Party took second place with 20% of the vote. After the election, de Klerk became a Deputy President in Mandela's ANC-led coalition, the *Government of National Unity*. In 1993 Mandela was awarded the Nobel Peace Prize. He shared the prize with de Klerk – for their peaceful termination of the apartheid regime. The two men stood shoulder to shoulder in Oslo.

Compare this with the situation in Myanmar as outlined above. Following her victory in the 2015 election she was offered, and accepted, the newly created role of "Head of Government". It was a title without authority. Nonetheless, it was because of her official role in the Government that she was criticised so heavily for not speaking out against the military crackdown in 2017.

While the Government's response is totally indefensible, there is clearly some truth in the claim that the crackdown was in response to attacks by Rohingya insurgents. Suu Kyi has devoted her life to a non-violent struggle for democracy and human rights. Does she hold the insurgents responsible for bringing the world crashing down on their own community just when democracy and human rights was within reach for all in Myanmar?

It seems that the majority of Buddhists support the expulsion of the Rohingya Muslims. It seems reasonable to assume that the majority of NLD members would also support their expulsion. This leaves Suu Kyi caught between a rock and a hard place. If she condemns the anti-Muslim campaign, she will lose the support of her party. If she does not, she loses the international support she won over the years. She knows that her battle to have a civilian government restored in Myanmar is far from ended. The Junta still has a vice-like grip on power. She needs a united party behind her if she is to break that grip. Time will tell if her silence will have a positive outcome in the long term.

UN System Was "Relatively Impotent"

Suu Kyi is not the only one to be criticised for failure to intervene in defence of the Rohingya. The UN is accused of ignoring warning signs of escalating violence before the alleged genocide. Early 2019, the UN Secretary-General, Antonio Guterres, appointed former Guatemalan foreign minister and UN ambassador, Gert Rosenthal, to look at UN involvement in Myanmar from 2010 to 2018.

The hard-hitting internal, report, released at the end of June 2019, condemned the organisation's "obviously dysfunctional performance" over the past decade and concluded there was a "systemic failure". There was no common strategy, even at the highest level of the organisation. As a result, the UN system was "relatively impotent to effectively work with the authorities of Myanmar. …Without question serious errors were committed and opportunities were lost in the UN system following a fragmented strategy rather than a common plan of action".

Rosenthal says that senior UN officials in New York could not agree on whether to take a more robust public approach with Burma or pursue quiet diplomacy and that conflicting reports on the situation were also sent to UN headquarters from the field.

The report emphasises the damaging impact of competing strategies between some UN agencies and individuals. Polarised approaches between quiet diplomacy with the Myanmar government and public condemnation of human rights abuses became more magnified as the situation in Rakhine worsened.

The report describes how the UN situation descended into 'unseemly fighting' where those that promote constructive engagement sometimes incur the wrath of those who favour a more robust advocacy role, and vice versa. One can only speculate that [former] Secretary General Ban Ki-moon was either unwilling or unable to arbitrate a common stance between these two competing perspectives.

The United Nations struggled to balance supporting the Burmese government with development and humanitarian assistance, while also calling out the authorities over accusations of human rights violations, Rosenthal concluded.

"The United Nations system…has been relatively impotent to effectively work with the authorities of Myanmar to reverse the negative trends in the area of human rights and consolidate the positive trends in other areas," he said.

"The United Nations' collective membership, represented by the Security Council, bears part of that responsibility, by not providing enough support to the secretariat when such backing was and continues to be essential," Rosenthal wrote. The 15-member Security Council, which visited Rakhine state in 2018, has been deadlocked, with Burmese allies China and Russia pitted against western states over how to deal with the situation. Rosenthal concluded that "The overall responsibility was of a collective character; in other words, it truly can be characterised as a systemic failure of the United Nations".

We know that Suu Kyi would have been on the side of quiet diplomacy and constructive engagement with the Junta. It seems reasonable to assume that she would look to the UN diplomats to "balance supporting the Burmese government with development and humanitarian assistance, while also calling out the authorities over accusations of human rights violations".

Those of us who wondered why Suu Kyi wasn't receiving support from the UN now know from Rosenthal's report that the UN situation had descended into "unseemly fighting" so that no support could be expected. The Junta were happy to take advantage of the "relative impotence" of the UN and press on with their crackdown.

Hope for The Future

The future looks bleak for the Rohingya Muslims and for the restoration of democracy in Myanmar. But I see glimmers of light at the end of the tunnel.

I didn't have a lot of contact with ordinary people. But what little I did have supported my theory that people are the same the world over. We all have the same basic wants and needs. We *all* have a dream – a dream of love, security, enjoyment and hope for a better future.

I found the people I spoke to in Yangon to be genuinely hospitable, helpful, and especially honest. I felt safe going out on my own. I went into town several times to wander around the Bogyoke market looking for souvenirs. There was a bus from the hotel to the city centre but one day I decided to try the local taxis. I have been taken for a ride in a number of countries by taxi drivers who are all too ready to cheat passengers who are not too sure of the local currency. I had no such trouble in Yangon where I was quoted the exact price and given the correct change.

I was impressed by the faith shown by ordinary people in the Shwedagon Pagoda when I visited. I am convinced that they pray to their God for a better future for themselves and for their children.

I am hopeful that there will be a growing awareness amongst ordinary people that the Military have gone too far and that there will be increasing support for those Buddhist leaders who have spoken out against the anti-Muslim campaign.

Abbot Arriya Wuttha Bewuntha of Mandalay's Myawaddy Sayadaw monastery said: *"This is not the way Buddha taught. What the Buddha taught is that hatred is not good, because Buddha sees everyone as an equal being. The Buddha doesn't see people through religion"*.

I have fond memories of my visits to Myanmar. It is a beautiful country, so rich in natural resources, so much potential for tourism. I am still hopeful that Suu Kyi's non-violent struggle for democracy and human rights will bear fruit.

Nepal
2001

The International Youth Coordination Council of Nepal organised a Conference to be held from 4-6th November, 2001 at Tigertops, Meghauli, which is situated in the Chitwan Province of southern Nepal. The title of the conference was *"Youth, Leadership and Development – Challenges and opportunities in the 21st Century"*. The group invited me to represent the IAYFJM and to present a paper on *The Convention on the Rights of the Child*.

The name Kathmandu had a special resonance for me. It echoed like a spell evoking images on a par with Kilimanjaro, Timbuktu or Machu Picchu – remote, inaccessible, mystical. It signified somewhere special.

I thought of Kathmandu as being at the foot of Everest. I had somehow expected it to be something along the lines of one of the Swiss Alpine villages with lots of snow and beautiful chalets. Instead, I found a city with (at that time) a population of over 1 million people. There were lots of old cars, old buses, old lorries all belching out smoke. The air was blue with fumes and filled with dust. It was choking to breathe, so I tried to breathe only through my nose. There were hundreds of three-wheeler motorcycles which had been converted into vans, taxis and the like. They were belching out smoke with a slightly different smell because of their two-stroke engines. It just added to the pollution.

I had waited so long to see Kathmandu. It was a terrible disappointment. Instead of a pristine little town nestling in the foothills of the Himalayas I found an industrialised city – overcrowded, polluted, run down. I wasn't even able to see Everest, let alone walk on the lower slopes. The myth was shattered. I was glad to get out of it.

I travelled 160 kilometres south by bus to Chitwan, a city of some 200,000 inhabitants. Most of the things I have said about Kathmandu would fit Chitwan also. I was met by two of the conference organisers who took me the final 20 kilometres by jeep to Meghauli where the conference was being held.

I was given an apartment in the magnificent Tigertops hotel. The hotel was surrounded by beautiful gardens with lawns and shrubs, which, in turn, were surrounded by trees and the river beyond.

The opening ceremony of the Nepalese conference was spectacular with a parade of some 3000 children led by elephants through the centre of Chitwan. We had the Deputy Prime Minister present, together with five other Ministers, so there was a high level of security. We had a police officer mounted on an elephant at either end of the stage where they had a clear view of the crowd. I was deemed to be one of the dignitaries and was driven directly to the stage where I had to sit cross-legged on a cushion. This meant I was not able to see the parade properly or to get photographs.

There was always an air of romance for me in stories of elephant safaris in far-off lands. I have an intense dislike of tiger-hunting which ends in the slaughter of those beautiful animals. But I could imagine myself on the back of an elephant moving slowly through the tall elephant grass, where the only shooting was done with cameras, or slowly meandering around ancient ruins. I hoped an opportunity might arise on my visit to Nepal.

The organisers promised that we would have the opportunity for an elephant ride before departure. We were right beside Chitwan National Park so I hoped we would go there. It was estimated at the time that there were some 300-350 one-horned rhinoceroses in the park, 60-70 Bengal Tigers, leopards, lynx, Bengali cats, jackals, foxes, wild dogs, civets, martens, mongooses, otters, sloth bear, wild boar, a range of monkeys and deer, gaur (wild cow) and some 450 species of bird.

I had hoped to go on a tiger safari in Chitwan National Park but "the opportunity for an elephant ride before departure" didn't suggest a tiger safari to me. I did however have several walks in the wood on the outskirts of the park and had a close encounter with a wild boar one morning. I heard the clatter of feet coming fast down the path towards me. As it got closer it veered off into the dense undergrowth. Wild boars are ferocious animals so I would not have argued with it if it had decided to stick to the path. I can only imagine that something was chasing it when it was in such a hurry. Later on I found a fresh footprint down by the river, which, judging by its size, could only have been made by a tiger. Perhaps it was the tiger which startled the boar. The staff in the hotel told me that tigers had been seen in this area. The footprint was all I saw. I also saw rhino footprints by the river and a heap of fresh dung – but no sign of the rhinos.

The forest had other less pleasant wildlife surprises for me. My mind was on tigers, wild boars and rhinos when suddenly I felt something on the back of my neck. I should have anticipated insect bites, covered up and used insect repellent. I hadn't thought of it. Now I was having some difficulty dislodging this thing. The blood was pouring down my neck and I suddenly realised that a leech had got me. They are not aquatic in this part of the world, living in the leaf litter instead. They find their prey by odour and sound vibrations. This one had dropped from the leaves above. It is not recommended to simply pull them off but I felt I had no other option. I didn't know that, if I left him alone, he would drop off himself when he finished feeding! I returned to the hotel where one of the staff managed to stem the bleeding and cover the wound with a plaster. I had no ill effects.

The leeches didn't keep me out of the forest. One morning I followed the sound of bells ringing and found a little Hindu temple. People come here early in the morning to pray. I was impressed by their faith with even young people there at 6 am.

As we approached the end of the conference, I didn't have much hope that the conference organisers would keep their promise of an "opportunity for an elephant ride". And then, on the morning of our departure, a number of elephants arrived and lined up beside a tall platform. We climbed the steps onto the platform and clambered onto the elephant's back, four to each elephant (plus the mahout[26]) sitting two either side, in specially adapted seats.

I had read so many stories about elephant safaris and now here we were on the real thing, well, almost the real thing, since we were heading back rather than out on safari. The elephants picked their way carefully through the elephant grass, which was as tall and sometimes taller than the elephants. It took us about half an hour to reach the river. I expected us to dismount there but the elephants took us on across the river and brought us to where the pickup trucks were waiting. It was a good experience and made up for the disappointment of not seeing any of the exotic animals.

In my week away from Kathmandu I grew to like Nepal and its people as they were and not as I had wanted them to be. As I rode along through the

---

[26] A mahout is an elephant rider, trainer, or keeper.

crowded streets once again, I felt part of the hustle and bustle, I felt the energy of the people. I felt that the pollution, which surrounded me, was not something they could easily do anything about. In the developed world pollution is a reflection of greed. Here it was a reflection of a people caught in the poverty trap who can't afford to buy new vehicles, or to take the necessary steps to avoid pollution.

As I headed for the airport, I thought it strange how much my attitude to Nepal had turned full circle in one short week. I arrived in Nepal, barely able to wait to set foot in Kathmandu. One day later I was glad to be leaving it behind as I headed for Chitwan. In the week since then I had developed a deep love for Nepal, for its people and for noisy, polluted Kathmandu. I hoped that, some day, I would have the opportunity to return.

Northern Ireland
1995, 2006

The IAYFJM World Congress, 1994, was held in Bremen, Germany in August of that year. I was elected to the 16-member Executive Committee.

Because the World Congress is only held once every four years the Council always encourages members of the Executive Committee to organise regional conferences in the intervening years. I decided to organise an international conference in Belfast in April 1995. 1995 had been designated by the UN as an International Youth Year, held to focus attention on issues of concern to and relating to youth. So, it was appropriate that the conference should take the violence of 25 years as its theme and endeavour to acquire a better understanding of the effect of that violence on our young people. By doing so, lessons could be learned which hopefully would impact on education, training and youth policy not only in Northern Ireland, but in many countries throughout the world. I called the conference *"Growing Through Conflict: The impact of 25 years of violence on young people growing up in Northern Ireland"*. I decided to limit attendance to approximately 100 people in order to be able to focus on the topic and have some meaningful discussions. I put my proposal to the Executive Committee at a meeting of the IAYFJM in Madrid in March 1993 and won their support.

It was hard work. I estimated that I would need to raise approximately £100,000 in sponsorship. And, of course, there was the small matter of *The Troubles* and the media images which had gone round the world of Belfast in flames with bombs going off and people being shot on a daily basis. The Americans were the most difficult. I pointed out that more people were murdered in New York every year than had been killed in Northern Ireland in all the years of the troubles up to then. But they were not listening.

I bombarded Association members with good news stories about Belfast. I told them how Cave Hill overlooks the city of Belfast and lies just a few miles from the city centre. All of Belfast can be seen from its peak, as can the Isle of Man and Scotland on clear days. It is distinguished by its "Napoleon's Nose", a basaltic outcrop which resembles the profile of the emperor Napoleon. It is thought to be the inspiration for Jonathan Swift's *Gulliver's Travels.* Swift imagined that the Cave Hill resembled the shape of a sleeping giant safeguarding the city. I told potential delegates that we would take them to see the world-famous Giants' Causeway, the Glens of Antrim, the Mountains of Mourne and the Lakes of Fermanagh. We would visit Bushmills Distillery and Belleek Pottery. They would get to walk on the Walls of Derry. I warned them that I planned to work them hard but I would be setting aside plenty of time to relax. Then I had a lucky break.

On August 31, 1994, which, coincidentally, was my birthday, the IRA declared a "complete" ceasefire. The IRA statement said that, after a quarter of a century of "armed struggle" to get the British out of Northern Ireland, there would be a "complete cessation of military operations" from midnight that night to allow inclusive talks on the political future of the province. I couldn't believe my luck. My colleagues in the IAYFJM, particularly the Americans, thought I had great powers of persuasion! I just prayed that the ceasefire would hold at least until after the week of April 3-7, 1995.

There was so much support for the conference in Northern Ireland that I raised the £100,000 without much difficulty[27]. We had 100 applications drawn from Argentina, Austria, Belgium, Brazil, Canada, Chile, England, France, Germany, Holland, Ireland, Italy, Japan, Kenya, Northern Ireland, Peru, Portugal, Scotland, Spain, Switzerland, Uruguay, the USA and Venezuela. Thus, we had representatives from all five continents.

The police were not convinced that the ceasefire would hold so we had lengthy discussions on the safest place to accommodate 100 judges from around the world and on a suitable venue for the conference. The police insisted that we could not hold it in the city centre. Judges, they believed, are impossible to control. They won't listen to advice. They would wander off and get themselves

---

[27] The conference was sponsored by: The Northern Ireland Office, The Northern Ireland Court Service, The Central Community Relations Unit, The European Commission – DG1A/A.2, The Department of Health and Social Services, The Department of Education for Northern Ireland (Youth Service Section), The British Council in Northern Ireland, The Northern Ireland Voluntary Trust, The Northern Ireland Tourist Board, The Law Society, The Bar Council, The BPS NI.

into trouble – not like our own judges who are used to following police instructions in the interests of security. We eventually settled on the Stormont Hotel – about three miles from the city centre.

The conference went off without a hitch. But I don't intend to discuss the conference here. I warned the delegates that they would be expected to work hard but I promised that I would set aside time to see the best of Northern Ireland.

During the conference, we had simultaneous translation in our three official languages – English, French and Spanish. Clearly this was not available once we left the conference centre. I was delighted, therefore, when my son Liam agreed to make himself available as an interpreter at all social events. This was invaluable as he could deal with questions in a range of languages, and not just with English, French and Spanish.

On Monday April 3, the Secretary of State for Northern Ireland, Sir Patrick Mayhew, held a reception and dinner in Hillsborough Castle. He highlighted the more positive side of life in Northern Ireland. He said he hoped delegates would take away the impression that life here was altogether different from that which they perceived before they arrived in the province.

On Tuesday, we were in Carrickfergus Castle. The evening began with a tour of the Castle followed by a display of Irish Dancing by a group of local children. Dinner was served in the Banquet Hall. Background music was supplied by an Irish Harpist, a "Fiddler" (playing traditional Irish music on a violin) and a vocalist.

On Wednesday, we had a Civic Dinner in Belfast City Hall hosted by the Lord Mayor and Belfast City Council. The Lord Mayor dispensed presents to all our guests.

On Thursday afternoon, we set off on a tour of the Antrim Coast – one of the most beautiful scenic routes in the world. We visited Old Bushmills Distillery, famous around the world for its Irish Whiskey, before going to see the Giant's Causeway, one of the Wonders of the World. We stopped off at the Centre for the Study of Conflict at the University of Ulster in Coleraine to hear a talk by Professor Seamus Dunn in which he discussed the nature of the conflict in Northern Ireland. Then it was off to the Ramore Restaurant for dinner. Situated on the stunning causeway coast in the seaside town of Portrush, the Ramore is renowned for its fantastic-tasting fresh food and lively atmosphere. With magnificent views over the harbour, this is one of Northern Ireland's most popular dining destinations. Informality goes hand in hand with quality and value – an excellent place to relax after four days' conferencing.

Friday provided an opportunity for our guest judges to get an insight into our court system. The morning started with a visit to the Royal Courts of Justice followed by coffee with the Lord Chief Justice. After coffee, one of the Appeal Court Judges gave a talk on the Family Court. Then the judges split into small groups to visit a court of their choice. The morning concluded with a buffet lunch in the Judges' Assembly Room. The afternoon was free for shopping.

On Saturday morning, we set off for a tour of the Fermanagh Lakes, then on to Belleek Pottery where I had arranged a surprise present for everyone who had stuck the pace so far.

There was an air of excitement as we entered the City of Derry because I had promised them the opportunity to walk on the famous Walls. Finally, there was time for a tour of the Guild Hall before a Civic Dinner and Reception, hosted by the Lord Mayor and City Council. This brought one of the funniest incidents of the entire week. When the dinner was over the Lord Mayor stood up to speak. He forgot that he still had his napkin tucked into the top of his trousers. The Canadian judge, Lucien, was sitting with me at the top table. The Lord Mayor's wife, who was at the other end of the table sent a message to Lucien's wife Joan asking her to remove the napkin. Joan made a grab for the napkin. She missed but had a second, and a third, attempt. By this time all the guests were falling about laughing. Meanwhile, the Lord Mayor, still unaware of his napkin and fearing he was the subject of a sexual assault by a sex crazy woman from Canada, had stepped back out of her reach. We deeply regretted that the incident had not been caught on camera. Today it would go viral on YouTube.

I had promised our guests one final tour. Many of the judges had already left for home when, on Sunday morning, we set off for the Mountains of Mourne. We visited Downpatrick, Dundrum Castle, Newcastle, the Silent Valley Reservoir, the Spelga Dam and Tullymore Forest Park.

We were all rather sad on Monday April 10, as the last of our guests said farewell. Many of our guests came with little knowledge of Northern Ireland beyond its image as a war zone. They left with fond memories of the warmth of our hospitality.

My reputation, high at the beginning of the week, was even higher at the end. We had glorious weather, with sunshine and blue skies and temperatures in the low 20°s C. I seemed to our guests to have influence in all the right places. Not only had I arranged a ceasefire with the IRA, I had persuaded God to give us a glorious week with a cloudless blue sky and warm sunshine every day!

The years that followed were fairly hectic ones for me with little time to even consider organising another conference. However, I did have discussions with the Director of the Northern Ireland Court Service about hosting the 2006 World Congress in Belfast to mark the end of my term as President.

The IAYFJM accepted our bid to host its XVII World Congress. The Congress was scheduled for August 27 to September 1, 2006 and would be held in the Waterfront Hall, Belfast. It was decided that the aim of the Congress would be to promote fresh initiatives internationally for the protection of children's rights and the progress of youth justice. The primary objective was to draw up a set of recommendations, which would serve as an inspiration to policy makers, professionals and judges throughout the world in the formulation, development and application of youth and family justice.

The title was: "The Right Justice: Putting the Pieces Together Again". "*The Right Justice*" had a double meaning. First it implied the "correct" justice – justice that would have a positive impact on dysfunctional families. Secondly, it implied the right of the child to family life. "*Putting the Pieces Together Again*" focused on the role of the courts in trying to mend broken families.

With more than 600 delegates, social events could not be on the same scale as for the conference in 1995 when we had 100 delegates. Events were organised in the Royal Courts of Justice, Hillsborough Castle, Queen's University Belfast and in Belfast City Hall. The congress dinner was held in the Europa Hotel – named by the media as "the most bombed hotel in the world"[28].

I called the Association's General Assembly for the afternoon of August 31. This included the election of the office holders for the next four years. At the end of the General Assembly, I handed over the Presidential Chain of Office to the incoming President. Then I urged delegates to hurry round to the City Hall where the Lord Mayor was waiting to greet them. I thought that would be a nice surprise for them. I was the one in for a surprise. When I walked in, I was greeted with a chorus of "Happy Birthday to you", a monstrous birthday cake and the biggest birthday card I had ever seen. A group of Maori policemen did a traditional dance

---

[28] The Europa Hotel is a four-star hotel in Great Victoria Street, Belfast, Northern Ireland. Opened in 1971, it has hosted presidents, prime ministers and celebrities. Between 1970 and 1994, it was damaged by explosions 33 times, gaining it the dubious title of "the most bombed hotel in the world".

as they came up to me to present the card! When I was busy organising the conference, my sister Una had been busy organising the birthday party. She did a good job. I was taken totally by surprise.

Quite a number of people were keen to invite me out to dinner. We only had one free evening, and there were thirty or forty of them. If I had gone out with one, a lot of others would have been disappointed. So, I came up with a compromise. Why didn't we all go out together for dinner? I booked tables in Benedict's. With a smallish group, we had plenty of time to chat. We had a great night's craic[29].

My son's interpretive skills proved invaluable once again, not just for the social events but also in drafting the recommendations, since submissions from the various groups came in a range of languages and not just the three official ones. He had to leave my birthday party early and spent most of the last night translating the recommendations and making sure the full text was available in all three languages! They were all printed in time for the final discussion on September 1.

Gerry, the conference secretary, wanted to carry out a proper analysis of the event. He distributed evaluation forms. 100% of respondents were very satisfied with the overall Congress experience – the organisation, the venue, the events, the exhibitions, the catering. 97% were satisfied with the website. 98% were satisfied with the registration process. The comments received indicate that an exceptionally high standard was achieved.

One particularly rewarding aspect of the evaluation was that the delegates all spoke about the warmth, friendliness and courtesy of everyone they met in Belfast. They had heard about "Ireland of the welcomes". Now they felt they had experienced it.

---

[29] "craic" is an Irish word meaning fun, entertainment, and enjoyable conversation.

Palestine
2005

In June, 2005, I attended an international conference in Bethlehem (Palestine) entitled "Kids Behind Bars – A Child Rights Perspective". I was particularly pleased to be able to visit the birthplace of Christ. I found it a moving experience to pray in the Church of the Nativity.

I was also moved as I took a short tour of the area of Palestine surrounding the city and witnessed the encroachment by Jewish settlers on Palestinian land. It was heart-breaking to be told that the houses surrounding almost every patch of greenery in the desert (the greenery indicating the presence of water) belonged to Jewish settlers. It was disturbing to see roads reserved for Jewish settlers while Palestinians had separate roads – a clear indication of apartheid. In some places, Israel's separation barrier cuts deep into the occupied West Bank, excluding Palestinian communities and annexing land around illegal Israeli settlements. The wall ran close to the hotel I was staying in so that I was able to walk up to it. I also watched as children threw stones at the wall. The soldiers on the top were in no danger but they still retaliated with CS gas.

I wanted to do *The Way of The Cross* in Jerusalem only to find that I could not take a taxi from Bethlehem. I could have taken a taxi to the wall (which was not necessary since I was so close to it) but once I passed through the barrier, I would have to take an Israeli taxi, or a bus, the rest of the way. I decided to take an old ramshackled bus, normally used only by Palestinians.

I enjoyed my *Way of the Cross* and what I saw of Jerusalem. The time passed very quickly and I was soon back on the bus for the return to Bethlehem. We were stopped by an Israeli patrol close to the wall and everyone on the bus, except me, was searched. The soldiers, who all looked like teenagers, picked on an old lady who must have been well into her eighties. She was taken off the bus and searched again. She was arguing with the soldiers but I couldn't understand what was being said. She produced a piece of paper which I assumed was her pass. The patrol leader took the pass, tore it up into little pieces and threw it at her. I couldn't imagine why he had picked on the old lady unless he was trying to provoke the men on the bus into retaliating. The problem for her was that she wouldn't be able to go to Jerusalem again without a pass.

As I watched this episode unfurl an image flashed through my mind of a visibly reluctant Yitzhak Rabin, under pressure from Bill Clinton and the world's media after signing *"the Declaration of Principles on Palestinian Self-Rule"*[30] on the White House lawn in September 1993 consenting to shake Yasser Arafat's hand. George W Bush did nothing to nurture this fragile peace process during his two terms of office. The incident I had just observed demonstrated that, 12 years after the Accord was signed, nothing had changed at ground level.

Benjamin Netanyahu was elected Israeli prime minister in 2009. He endorsed the two-state solution in a speech that year, but he continued to expand West Bank settlements and, in 2015, said there would be "no withdrawals" and "no concessions." President Trump hammered yet another nail into the coffin of the Middle East peace process on December 6, 2017. In recognising Jerusalem as Israel's capital, he demolished 70 years of US policy.

---

[30] The Oslo Accords.

Peru
2009

The Inca Civilization

The Inca civilization was the largest and most powerful state ever to occupy the South American continent. Its glorious and majestic past has long had a fascination for me. Starting with the foundation of Cusco around 1250 AD it grew in strength and importance over the next 200 years. During a period of rapid expansion, which began around 1438, the Incas conquered most of South America. They conquered hundreds of kingdoms covering thousands of square miles and turned them into one vast multinational state, achieving incredible levels of development.

The level of social and political organisation, the knowledge of architecture, astronomy and engineering and the outpouring of textile art and ceramics underscored its claim to be one of the most important cultures of the ancient world.

Their understanding of agriculture was far in advance of their time. In a valley near Machu Picchu, they developed a series of terraces, each with its own micro-climate where they were able to experiment with various crops in order to develop varieties which were more suitable to local conditions.

And yet this blossoming of Inca culture lasted only 100 years. Their empire came to an end in 1533 with the execution of the last Inca (ruler), Atahualpa, by the Spanish conquistadors.

The Inca civilisation has left us a rich heritage: food such as potatoes, tomatoes, maize and quinoa. The Incas also left us drugs such as quinine and tobacco, and an amazing network of paved trails punctuated by superb ruins. Quechua, the official Inca language, is still spoken by 10 million people to this day.

Andean people maintain elements of the Inca religion, such as making offerings to *opus*, the spirits of the sacred mountains, while also practising the Catholic religion. Our guides, when having a drink, would offer some to "mother earth" before drinking themselves. Above all, the Inca values of hard work and cooperation are obvious in the lives of their descendants.

The fact that comparatively little is known about the Incas is because the Spanish set out to destroy all evidence of their culture and civilization, even plundering the tombs of the Inca nobility and priests, looting the gold and silver ornaments encrusted with turquoise and other precious stones, before eradicating all traces of the tombs. They melted down virtually all Inca gold and silver in an appalling display of greed and vandalism.

The Incas constructed a vast road system which connected the four ends of the empire to Cusco and the towns and cities with one another. It has been calculated the road system extended to 30,000 kilometres.

One of the most famous of the Inca trails was the one which follows the Urubamba River for about 50 kilometres from Ollantaytambo, through the Sacred Valley, to Machu Picchu. The highest crossing point of the trail reaches an altitude of 4,200 metres.

The Incas assigned a sacred character to this valley because their astronomers and priests believed that it was a projection of the Milky Way. Fortunately, the Sacred Valley lay hidden behind the mist of a humid forest of exuberant vegetation, was not found by the Spaniards and hence was spared destruction. It was not discovered until North American explorer Hiram Bingham stumbled across it in 1911. It has been open to trekkers since 1941.

Machu Picchu

Trekking the Inca Trail.

It had always been my ambition to follow in the footsteps of the Incas through the Sacred Valley to the mystical Machu Picchu. On two occasions in the past, I received an invitation to a conference in Peru but on both occasions pressures of work meant I was unable to attend. Then, in 2009, I received an invitation to attend the 1st World Congress on Restorative Juvenile Justice[31] which was to be held in Lima from November 4-7. This time I was able to accept the invitation. I asked Una if she would like to come with me with a view to trekking the Inca Trail.

Una and I arrived in Lima just before midnight on November 28 and took a flight to Cusco a few hours later. Cusco, which is close to the starting point of our trek, is at an altitude of 3,200 metres. We were both feeling the effects of the altitude soon after arrival – headache, nausea, dizziness, shortness of breath and, for me, a strange tingling sensation. I didn't sleep much that night, but there was no time for resting. We were picked up next morning at 6 am and taken by coach to the start of the Inca Trail to begin our four-day hike to Machu Picchu. We were joining a group of 15 others. I was the eldest at 71+. There was a man of 65 and his wife aged 60; then Una at 42; the others were all in their 20s and 30s.

The Trail goes up through the mountains, reaching 4,200 metres at its highest point before descending to Machu Picchu. We had reached a height of 4,600 metres on Kilimanjaro, so we thought we would be OK. We expected the porters to carry 12 kilos of our equipment (as the porters on Kili had done) so that we would only have to carry water and a few personal items. However, the chief guide told us the porters would only carry 7 kilos and we would have to carry the rest. So, instead of carrying 5 or 6 kilos in my backpack I had 10 or 12 kilos. We had been told to expect the temperature to be 'cold' with the possibility of rain. In fact, the temperature quickly rose to around 30°C!

It was an inauspicious start to the trek – already suffering from altitude sickness, carrying a backpack which was much too heavy in sweltering heat. A long haul up a very steep incline (about 70°) almost finished me off. One of the guides offered to carry my backpack for the final stretch to ensure that I reached the campsite.

On reaching the camp site I took a severe fit of shivering and could hardly manage to get a jacket on. I decided to skip tea and crawled into my tent to lie down for about 90 minutes. I decided to try to eat something when called for

---

[31] Restorative Justice promotes the idea that because crime hurts, justice should heal. Victims take an active role in the process. The victim is asked to tell how he/she has been affected by an injustice and to say what should be done to repair the harm. Offenders are encouraged to take responsibility for their actions and repair the harm they've done—for example by apologising, returning stolen money, fixing the broken window or doing community service. Help is provided for the offender in order to avoid future offences.

dinner but could only manage some soup and skipped the main course. One of the other trekkers offered me some tablets for altitude sickness – one to be taken every eight hours.

Next morning, we were up at 4:30 am to begin packing. I still didn't feel like eating and had only two small pieces of French toast for breakfast. I took one of the altitude sickness tablets and gave the other one to Una.

The chief guide said he would get a porter to carry my backpack since I was still so much under the weather. We were on the road again about 6:20 am.

It was a relief not to have to carry the backpack but as the paths got steeper and the temperature rose my progress slackened to a crawl. After about four hours I sat down with Una to review progress. At this pace, I reckoned it would take me five hours to reach the summit plus another four hours down the other side to reach the second night's camp. I didn't think I had any chance of reaching that camp site. The only sensible thing would be to turn back. Una was already concerned about me. She said she had watched me getting slower and slower and paler and paler. She agreed we should turn back.

We waited for the chief guide. He had been expecting this. He asked Una if she would like to continue but she said no. He arranged for a guide and porter to go back with us.

We had already been trekking for four hours and two more hours took us back to the previous night's camp site at Huayllabamba. I expected we would camp there but the guide pointed out that there were absolutely no facilities. The porters go ahead of their group and set up the tents and cooking facilities for the night. But next morning, once the group leaves, the porters pack up everything and move on.

I was shocked when the guide said we would have to continue to KM82, the point where we had commenced the hike on day one. That took us five hours on the first day when I was reasonably fit. At this point I was totally shattered after hiking for some six hours. There was no way I could manage another five hours hiking.

There was no point in ringing for a taxi up here. There was only one option – we hired two horses. Neither of us had any experience of horse riding and this was hardly the place one would choose to begin training. Going up or down very steep inclines, many including steps, is testing terrain for even experienced riders. It reminded me of the Irishman giving directions – "If I wanted to get there, I wouldn't start from here". But here we were, high in the Andes, unable to walk any further but unable to stay where we were. There was no other option and no time for trial runs.

We mounted the horses and set off. Never in our wildest dreams had we imagined doing this but here we were trekking on horseback through some of the wildest terrain imaginable. At times the horses got much too close to the edge and, as I gazed into the abyss below, I wondered whether they would ever find my body should the horse slip. I wondered how Una was faring on her horse but did not dare to take my eyes off the trail ahead to look behind me. I needn't have worried. The horses are at home in these mountains and are much more surefooted than I was giving them credit for.

We only dismounted once – when we came to the very steep, 70°, incline since there was a serious danger that we would slide forward over the horse's head. And, of course, there was no headgear to protect our head should that happen. It was not a comfortable ride and this was a good opportunity to ease aching buttocks. But riding was better than walking so we remounted as soon as we got past this precipitous slope.

We surprised ourselves at how well we managed and we eventually reached the beginning of the trail after about 2 hours 50 minutes on horseback. We were saddle sore and weary but relieved to have reached our objective.

If anyone had told us before we left home that we would be trekking on horseback through the roughest mountain terrain in the Andes we would have laughed at the very idea. And yet here we were having achieved what we would have believed to have been the impossible. It was not the target we had set ourselves but an achievement nonetheless.

We bade farewell to our horses and hired a taxi to the nearest town – Ollantaytambo where we booked into an hotel. It was so nice to have a hot shower and climb into a proper bed. I should have been so disappointed with myself having failed to achieve my ambition of walking to Machu Picchu. Instead, I was quite relieved – convinced that I had made the right decision.

Next morning, we sat in the garden watching the hummingbirds and thinking of our colleagues trekking over the mountain. I pondered why I had suffered from altitude sickness on this occasion when I had climbed higher in Kilimanjaro with no ill effects. The major problem I felt was that the travel agent had changed our schedule. Our original plan was to attend my conference first and then trek the Inca Trail. A change of schedule meant that we had to do the trek first and there was no time to get used to the altitude. We had been told to expect cold, damp weather. The temperature soared to 30ºC. There was also the problem of porters refusing to carry our baggage – something we had not been told in advance. I might have survived if someone had carried the bulk of my things from the beginning. I was not as fit as I had been for Kili. I was not aware until I got back home that I had major heart problems and would need a quintuple bypass. According to the surgeon, my arteries were the worst he had ever seen. Little wonder then that I was struggling on the Inca Trail. But I was unaware of my heart problems at the time.

We caught a train which took us through the amazing Sacred Valley with the raging torrent of the Urubamba River on one side and sheer cliffs rising hundreds of feet on either side of the valley. We arrived in the town of Machu Picchu and booked into an hotel. We spent the afternoon wandering around the markets and buying souvenirs.

Next morning, we took the bus to Machu Picchu itself where we planned to meet up with our erstwhile colleagues who were due to complete their trek that morning. By 8:30 am they had all arrived at the assembly point, tired and weary, but elated that they had completed the trek.

Machu Picchu was all I had ever imagined it to be – one of the most important cultures of the ancient world. Inca architecture shows an almost religious commitment to suiting the design and choice of materials to the site. Inca stonework is famous for its mortarless walls with large stones fitted together with amazing precision. Nowhere throughout the entire site did I see grass or plants growing in the gaps between the stones, which gives an indication of how tight the fit was. Moving the great stones into place needed great manpower. It is believed that they used rollers of wood and stone.

Inca religion featured worship of the sun, moon, and stars. Having trekked up steep mountain paths in the Andes under a blazing sun and temperatures of more than 30°C I could not share their enthusiasm for the sun god. However, it is easy to understand worship of the sun if one remembers the bone-chilling cold of a clear winter's night in the Andes.

At sunrise on the winter solstice (June 21 in the Southern Hemisphere) a shaft of light shines through one of the windows in the Temple of the Sun and aligns perfectly with a slot which the Incas cut into a large rock. A second window permits the sun's rays to enter during the summer solstice (December 21). A third window is known as the Window of the Serpents. It has perforations around its frame through which, it is believed, serpents were introduced into the room, presumably involving animal sacrifices.

A minute or so uphill from here along an elaborately carved stone stairway brings you to one of the jewels of the site, the Intihuatana, which means "the hitching post of the sun". It is thought by some to be a symbolic representation of the spirit of the mountain on which Machu Picchu is built. By all accounts it was a very powerful spot both in terms of sacred geography and its astrological function. People were queuing up to touch the rock to feel the energy it was transmitting.

Sophisticated ceremonies and rituals were held at the rock at various times of the year. For example, during the winter solstice the Incas asked the Sun God not to desert them but to start a new cycle introducing days which would gradually get longer and warmer. In some ways this represented their attitude to death which they did not see as the end of things but rather the spirit passing to another world where things would be better.

I could go on about the Temple of the Condor, the Royal Plaza, the Royal Tomb – too many wonders to take in at one visit. Hopefully I have given you a flavour of what an intriguing and mystical place this is. We toured the site for a couple of hours and then took the bus back into town. Later that afternoon we took the train back to Cusco and the following morning flew back to Lima in time for my conference.

The conference was fairly hectic. Lima is a city of some 11 million people and our hotel was over an hour's drive from the convention centre. We were up at 6 am for breakfast and on the bus by 7:30 am, not getting back to the hotel until around 7 or 8 pm each evening. Luckily a friend of mine from South Africa had brought her husband with her and he and Una were able to explore the city each day while we were away working!

After the conference, Una and I spent a few days in Paracas National Park – some 3½ hours by coach south of Lima. We had a beautiful hotel with a fine swimming pool. We spent most of our time relaxing and birdwatching. We fell for the beautiful Inca Tern, which became one of our favourite birds. We went on a bus tour and saw boobies, pelicans, flamingos, oyster catchers, skimmers, swallows. We took a boat trip to the Ballestas Islands and saw sea lions, dolphins, penguins, two types of cormorants, Inca terns and boobies.

We had two days back in Lima before catching our flight home which allowed us time to explore the city together. We loved St Martin Square (my house in Belfast is called "St Martin") and San Francisco Cathedral. A police officer gave us a map and some advice on where not to go in the city. We took her advice and had no problems.

We left Lima on Friday November 13 and finally got back home, via Toronto and London, on Sunday November 15.

I had barely time to unpack my case on returning from Lima when I got a request from the Council of Europe to come to Strasburg to review the draft guidelines on *Child Friendly Justice*. 2009 had turned out to be busier than I had anticipated.

Poland

2003

Warsaw

I was invited to attend the IX European ISPCAN Conference on Child Abuse in Warsaw at the end of August 2003. The theme of the Conference was "Promoting Interdisciplinary Approaches to Child Protection". I had never been to Poland and was delighted to accept. I would only be in Warsaw for three days and most of my time would be taken up with the conference. There wouldn't be much time for sightseeing.

I wanted to see the Old Town, even though it was "old" only in name. The capital's historic centre, dating back to the 13th century, was all but obliterated by the Germans, who occupied the city early in the Second World War. Hitler saw Warsaw as no more than a railway hub and wanted it wiped off the face of the planet. After the Warsaw Uprising was brutally quashed, the Nazis began the systematic destruction of the city, building by building, street by street. 80% of the city was levelled.

Warsaw is sometimes called the Phoenix City because of the way it rose again from the dust. Gothic churches, neoclassical palaces, modern skyscrapers and Soviet-era tower blocks are interspersed with restaurants, designer shops, pubs and bars. Strangely, this unusual mix of architecture sits harmoniously together and reflects the city's turbulent past. The Old Town was faithfully restored. Its heart is Market Square, with pastel buildings and open-air cafes. The Monument of the Warsaw Mermaid, at its centre, is the city's symbol. There are numerous stories of how the mermaid became the symbol of the city. I like the one which says that she lived with her sister in the Baltic Sea before they decided to go their separate ways. Her sister went to Copenhagen. She swam up the river to Warsaw where she became entangled in fishing nets. Thankful to the fishermen who cut her free she now stands in Market Square ready to defend the city and its residents.

I wanted to see "the Stalinist behemoth" – the towering Palace of Culture and Science. The building is a reminder of Warsaw's days under communist rule. It was designed by Soviet architect Lev Rudnev and has become known as the Eighth Sister, because of its likeness to Stalin's Seven Sisters in Moscow. Liam and I stayed in one of them – the Hotel Ukraina – during our visit to Russia in August 2000.

Finally, I went to see "the Warsaw Genuflection" – a plaque that commemorates the occasion in 1970 when the German Chancellor, Willy Brandt, visited the memorial to the Warsaw Ghetto Uprising. Brandt laid a wreath and then dropped to his knees and remained silent for some time reflecting on the more than 13,000 dead and almost 60,000 deported. I didn't drop to my knees. Instead, I went into a church nearby to reflect on man's inhumanity to man and to say a prayer for those who had suffered such appalling cruelty.

Russia
2000, 2004

Towards the end of 1999 I received a request from one of our members in Russia, Judge Oleg Osheev, asking for help in dealing with young offenders. The dilemma which the judge faced was that he had no alternatives to custody. He had two options. He could either, find the young person not guilty and release him (or her), or he could find them guilty and sentence them to a term in a rather gruesome prison. Could I help?

The judge lived in the town of Chaykovsky, about 800 miles east of Moscow, 400 miles from Saint Petersburg and about 200 miles southwest of Perm. The town was founded in 1955 as a settlement serving the construction of a huge hydroelectric power station in Votkinsk – the birthplace of Pyotr Tchaikovsky. The new town was named after the famous composer. With a population of around 80,000, it is located in the Cis-Ural region on the left bank of the Kama River, near its confluence with the Saygatka River. The confluence of the Kama and Saygatka Rivers and the nearby Votkinsk Reservoir form a peninsula, on which the town is situated. It is a famous centre for tourism in the Western Urals. The closest airport is located in Izhevsk, about 60 miles away. Most tourists come by ship.

I arranged to go early in 2000. I hadn't a word of Russian and can't even hazard a guess at place names because the Cyrillic script is so unlike ours. My son Liam is a linguist and I knew he was quite competent in Russian even though he refused to describe himself as "fluent" in the language. I asked him if he could take time off and join me for the trip. He could fly direct from Geneva and we could meet at Saint Petersburg airport. He had never been to Russia so was delighted at the excuse to go and have the opportunity to practice the language.

The first hurdle I had to overcome was getting a visa. It was usual to provide a letter from a travel agent or from the agency issuing the invitation confirming the purpose of the visit. However, my invitation was from a private individual and I was organising my own travel arrangements. The staff in the Russian Embassy in Dublin were very difficult. Eventually I appealed to the Ambassador who, with rather bad grace, I felt, issued me with a visa. He then undermined the whole purpose of the trip by including a note which said I was not allowed to travel beyond Moscow! I wasn't sure that the note had any legal validity. I took the visa and said nothing, working on the assumption that I could get the matter resolved in Moscow. In the event, no one even questioned my visa. I had unrestricted travel in Russia.

I met up with Liam in Saint Petersburg where we planned to spend a couple of days. Liam's help was invaluable in getting transport into the city and later going sightseeing. Saint Petersburg is a beautiful city, with so much to see. Unfortunately, we didn't have as much time as we would have liked. One half day was given over to meeting with the Appeal Court judge who took key responsibility for those judges in the lower courts who were dealing with young offenders. We had a very interesting discussion.

We flew to Izhevsk – the nearest airport to Chaykovsky. Two young solicitors, Vasily and Max, who worked with Judge Oleg, met us there and drove us the last 60 miles to our destination. Both were heavily involved in a range of programmes and projects aimed at keeping young people out of trouble. Liam was invaluable as an interpreter. Vasily could speak a little English so that if Liam was stuck for a word, he could help out.

Judge Oleg was a quiet, softspoken man, who was convinced that prison did more harm than good, particularly for young people. Vasily and Max had assembled a group of volunteers who were doing excellent work in keeping young people out of trouble. But a spell in prison did untold damage. Many young people who went to prison for comparatively minor offences came out as hardened criminals. The legislation left no room for manoeuvre. Could I come up with a solution?

I asked Oleg whether the police agreed with him that prison did more harm than good. He said they did. Would the police be prepared to "present no evidence at this time" where the offence was not too serious, allowing the judge to release the young person? Oleg asked what the purpose would be. I said that he could release the young person if the police presented no evidence. He could impose a condition that the young person attend a relevant programme for the next six to twelve months. A report would be prepared and presented to the judge.

If the report was positive the young person would be regarded as having paid his/her debt to society. I kept in touch with Oleg and Vasily for some years and they were happy with how the scheme was working.

It was easy to see why Chaykovsky was attractive to tourists. It was in a beautiful setting surrounded by water and forests. And the people were so friendly. Vasily invited us to his home for dinner and we had an opportunity to meet his wife, his family and his parents. His father built the house which he, Vasily, was living in,

Before dinner, Vasily said he wanted to introduce us to one of the oldest Russian traditions – the Russian Banya (sauna). He led us to a large wooden shed in the garden. Inside, there were wide wooden benches along the walls, built up one above the other like steps. He said we could sit or lie down on the benches. The banya was heated with firewood. The higher up the bench the hotter the air is. There was a bundle of twigs and leafy branches from a birch tree, bound together like a broom. Once we were warmed up well enough, we should take the "broom", dip it into a tub of cold water and then smack it briskly all over our body. Vasily said that, in the Winter, they would go out and roll in the snow.

It's considered that the banya atmosphere brings people closer together, allows them to communicate and interact on a more common level. Usually, people take a break from the hot temperature and relax, drink aroma tea or special herbal tea in an adjoining room where they share their ideas and beliefs or chat about life in general. We had something more substantial to look forward to. Vasily's wife was preparing dinner. We got dressed and went in for one of the best meals we had ever tasted.

Vasily asked if we would like to do a parachute jump. I declined because we couldn't risk an injury with so many important meetings pending in Moscow. He then arranged for us to be taken on a helicopter trip. Liam was thrilled. We got a magnificent bird's eye view of the whole area. Next, he and Max invited us to go hunting in the forest, but hunting is not our scene. So, they organised a picnic instead. We crossed the river in a boat (it is quite wide at this point) and set off into the forest foraging for mushrooms. Max said he was going to make us a stew like we had never tasted before. He showed us the kind of mushrooms he wanted and the four of us spent an hour or so foraging until Max said we had sufficient. The stew was made entirely of potatoes and mushrooms and tasted just superb. We were joined by Oleg and by Vasily's family. We had a pleasant afternoon sitting in an idyllic spot overlooking the river. What a wonderful way to round off our visit to Chaykovsky.

We flew 600 miles southwest to Moscow where I had arranged meetings with Anna Guertisk of BICE (Bureau International Catholique de l'Enfance) and the "Centre pour la réforme judiciaire et légale" which was working very hard to promote the introduction of juvenile courts in Russia. I promised that the International Association would offer whatever support it could to the campaign. I also met with Rosemary McCreery, UNICEF's country rep for Russia, who agreed to throw UNICEF's weight behind the introduction of juvenile courts.

Our time in Moscow was restricted with just enough time for a couple of day's sightseeing. We loved Red Square and St Basil's Cathedral in particular. We visited Lenin's tomb, and after queuing up to get in, had to run back into a nearby park to leave our cameras for safekeeping. We were impressed with the large GUM department store with its unique glass roofed design. We found most goods in GUM expensive – the same Western goods you see everywhere, at Western prices – so we didn't buy a lot – apart from a few souvenirs. But they did have some nice coffee shops. We visited one of the street markets where clothes were being sold at unbelievably low prices. We bought shirts costing £3 each, which are still wearable 21 years later!

I was back in Russia in May 2004 at the invitation of my friend Valentina Semenko, the Supreme Court judge who, at that time, was a member of the IAYFJM Executive Committee. A motion to set up a juvenile court system in Russia had just been defeated in the Duma (Russian Parliament). There was now a movement to set up a pilot juvenile court and it looked as if the President, Vladimir Putin, would support that initiative. Could I come and offer my support? That would be sufficient to swing the decision.

Once again, I called on my son Liam to come with me and act as interpreter. These would-be higher-level meetings than we had in Chaykovsky but his Russian had improved greatly in the intervening years and I was confident he could cope. I caught an early morning flight with Swiss Air to Geneva so that we could travel together to Moscow.

We were met at the airport and taken by limousine to our hotel. Next morning, we were picked up again and driven 155 miles northeast to Ivanovo. Ivanovo is the capital of the province of the same name in Western Russia. The first linen mills in Russia were founded near Ivanovo by order of *Peter the Great* in 1710. A large number of weaving mills and textile-printing factories were subsequently opened there, so that by the middle of the 19th century the town was known as "the Russian Manchester". It is also famous for being a good town for a young man to find a fiancée, since most of the workers in those industries were female! It reminded me more of Belfast, even in size, with a population of about 500,000. It remains one of the major textile cities of Russia.

We had several meetings with Valentina, who was Chair of the Regional Department of the Russian Children's Fund based in Ivanovo; with Uriy V Smirnov, President, Ivanovo Regional Court, with some of the senior judges; and with Igor E Gladkov, Chair, Arbitration Tribunal of Ivanovo Region, together with some other local politicians. Surprisingly, President Putin sent a representative to hear the evidence in favour of setting up a pilot juvenile court in Ivanovo. He seemed duly impressed and said he would make a positive recommendation.

Back in Moscow for a few days' sightseeing, we finally got to visit the Kremlin[32] after waiting in a long line! We were surprised to find three cathedrals and a couple of other churches there. One of the cathedrals held the tombs of the Tsars. In the Kremlin we also visited an exhibition of Fabergé eggs. On other days we went to the Tretiakov Art Gallery and the Pushkin Museum, and to the spectacular cathedral of Christ the Saviour which had been rebuilt at a cost of $200m and had opened just after our first visit in 2000. The original had been blown up by Stalin in the 1930s and replaced by an outdoor swimming pool. The humidity, from the pool, was believed to be damaging artworks in the nearby Pushkin Museum. We also spent a lot of time on the amazing Moscow metro. Stalin called the metro stations "Palaces of the People". They certainly are outstanding. They are the only metro stations I have been in where I wanted to spend time wandering around admiring the artwork!

---

[32] "Kremlin" originally meant "fortress".

Scotland
1986

In 1986, I was invited to be the NI Representative to the Annual Conference of the Reporters to Children's Panels (Scotland) which was held in the Crieff Hydro Hotel in the beautiful market town of Crieff. My invitation to the Reporters Conference was to become an annual event.

Crieff lies on the A85 road between Perth and Crianlarich. It is best known for its whiskey and its history of cattle droving. It has been a well-known tourist centre since the 19th Century – many attracted to the Crieff Hydro Hotel which opened in 1868. For me the main attraction, apart from the conference, of course, was Perthshire's stunning scenery and the opportunity to go birdwatching.

My colleagues back home were interested in my report from Scotland and were keen to know how the Scottish system differed from our own. Consequently, I organised and led a delegation from Northern Ireland to Scotland, in June 1987 to study their system.

The Children's Hearings System in Scotland

Pre-1971, children and young people in Scotland were dealt with by juvenile courts, whether they had committed an alleged offence or were in need of care and protection.

In 1960, a committee was established under Lord Kilbrandon to respond to concerns over youth justice and investigate possible changes to the approach.

In 1964 the committee reported that there were great similarities in the need for care of all children and young people appearing before the courts, regardless of their reason for being there. This report led to the creation of the Children's Hearings System.

The Children's Hearings System began operating on 15 April 1971[33], taking over from the courts the responsibility for dealing with children and young people who are in need of care or protection or who have committed alleged offences.

One of the fundamental principles of the new Scottish system was that children and young people who committed offences were to be dealt with in the same way as children and young people who needed care and protection. This was an argument I and a number of my colleagues had been making to the Northern Ireland Office for years. We were delighted that Lord Kilbrandon supported what we had been saying.

We had also argued for less formality in children's hearings. For example, no gowns to be worn, hearings should be in rooms with all parties sitting at the same level, preferably around a table. Kilbrandon had taken this a stage further:

"The Children's Hearings System was to include a panel; neither a court of law nor a local authority committee. The Panel was to be a lay body, comprising people from the local community who would make decisions with and for children and young people in their community."

My colleagues and I were wondering how the new system was working in practice. I managed to arrange that members of the delegation would have the opportunity to sit in on Children's Hearings as observers (not as a group, but as individuals – one per Hearing).

We were impressed with what we saw but had some concerns. Children aged 16 to 18, charged with an offence would still be prosecuted in a sheriff court or the High Court of Justiciary. Children under 16 could similarly be prosecuted where the procurator fiscal decided that the seriousness of the case merited it.

We were impressed with the informality. Everyone sat around a table and everyone was casually dressed. It seemed very relaxed to us but parents had a different perspective. One parent told a researcher: "Everyone had a big file in front of him/her except me. And everyone was taking notes, except me. Everyone was talking and asking questions of me and my son. My head was in a spin." We, too, had plenty to think about on the way home.

---

[33] There have been important changes since our visit in 1986. Those interested should go to
https://www.gov.scot/policies/child-protection/childrens-hearings/

South Africa

2002-2007

South Africa was one of my favourite destinations. I was lucky enough to get involved with a group of people dedicated to promoting the best interests of children. The group was made up of a Supreme Court Judge, university professors and lawyers. Over a period of six or seven years I attended annual seminars and did some work with the University of the Western Cape.

I visited Pretoria, where I presented a paper at the university, and Johannesburg, where I addressed the city's Magistrates. But mostly my visits were to Cape Town. I love Cape Town. I love the waterfront and I love Table Mountain – I was up there on every visit to see the beautiful birds, butterflies, lizards and "dassie". Dassie are properly called "rock hyrax". They are sometimes called "rock rabbits" or "rock badgers". I think they look more like overgrown guinea pigs – they can weigh up to 4 kg (just under 9 lbs). I went birdwatching along the coast east of the city centre. I knew where to find the black oyster catchers, which are on the endangered species list, and where the blacksmith plover nested.

Usually, I didn't have a lot of free time but, in 2006, I decided that there was a possibility I might not be back since I was coming to the end of my four-year term as President of the International Association of Youth and Family Judges and Magistrates. I didn't want to leave Africa without having seen at least some of its wild animals. I wanted, especially, to see "the Big Five" (elephant, leopard, rhino, lion and buffalo) and, perhaps, giraffes, hippos, cheetahs, crocodiles and hyenas. There is such an abundance of wild animals, not to mention many different species of birds that I didn't expect to be disappointed. So, I allowed time for a short safari. I won't go into too much detail on the safari at this point because I came back with Una the following year for a longer safari and there is much overlap between the two accounts. I do want to comment on two incidents which stand out in my mind. Thereafter the two reports are merged.

As we drove around the park on day 1, I spotted several lions feasting on a kill. I asked the guide what they were eating. He replied that they had killed a giraffe the previous day. I was surprised. I thought that, because giraffes are so big, they would be safe from predators. The guide replied that lions frequently kill giraffes. Lions have been known to attack and kill an elephant when no other prey was available.

The following morning, we set off on safari before dawn broke. We had just left our camp and were travelling slowly along a narrow track when we were joined by an uninvited guest. A hyena had emerged from the jungle and ambled along beside us – literally inches away. I was aware that hyenas usually live in clans and are generally nocturnal. What was this one doing out alone just before dawn?

Hyenas were high on the list of animals I wanted to see. But I wasn't expecting to have one so close. He looked quite harmless as he strode along beside us. I could have reached out and stroked him. But I was aware that hyenas have powerful jaws and can crunch and swallow the bones, and even the hoofs, of their prey. And this hyena was not laughing. Indeed, he seemed in a grumpy mood. He glanced up at us every once in a while and made a low growling noise. I interpreted that to mean "please go away and leave me in peace" although his demeanour suggested he had reduced that to two syllables. I guessed that, had I attempted to pet him, I would have been the one who was armless. Eventually he got fed up with us and ambled off into the jungle.

I never imagined Una would want to go on Safari as she can't stand "creepy crawlies". But when I told her about all the wonderful animals, and beautiful birds, I had seen, a South African safari shot to the top of her "must do" list. We booked a trip to South Africa from 19 January – 3 February 2007.

South African Safari

This was a holiday and not a working trip, and Una was coming with me. So, things were different to my usual "missions". We spent the night before packing our cases and sorting out all the things around the house that needed doing, such as watering plants and checking security. The morning of our departure we had to go to the kennels with Lucy (our little toy poodle). Una didn't want to leave her but she did want to go on the trip. Lucy began shaking as soon as we arrived at the kennels and looked so sad as we walked out the door and left her. We both found it hard to turn away and walk out without her.

We flew to London at 1 pm and had a long wait at Heathrow as our flight to Johannesburg didn't leave until 9 pm. However, we were able to relax in the business lounge which is much more comfortable than having to sit in the main waiting area. We took off on time and had a very uneventful 10½ hours.

Once we had picked up our cases in Johannesburg, we were surrounded by porters asking us where we were going. They came in quite useful as we had to leave the international terminal and walk to the domestic terminal to get the next flight. It was only about a 10-minute walk but we wouldn't have known where to go.

Our instructions were to check-in at desk 63 and when we got there, we were told it was the wrong desk. However, the staff were very helpful and one went to check out which desk we should have been at. When we got to the right desk Una was told her case was too big to take on the flight. Everything she needed was in the case. They talked about it for a while and then decided she must go and buy a bag and take only what she needed for the safari. The lady from the desk brought us to the shop and then down to the lock-up to leave the big case and then back to check-in. Our flight was due to leave in about 10 minutes so we went straight to security and down to the gate. The lady from check-in was already there waiting for us. We discovered that we were travelling on a seven-seater plane and that was why Una couldn't take her big case. We had a 1½-hour flight. We landed at a small dirt airstrip and there were a couple of jeeps waiting for us. We were the only two going to Chitwa Chitwa which was about 15 minutes' drive. This is a beautiful place. A private game reserve bordering Kruger National Park. It is set on the side of a lake and no matter where you are you have a view over the lake. We signed in and were taken straight to lunch while our bags were taken to the cabin. Lunch was served under a thatched roof sitting on the edge of the lake. We were joined by a couple of black-faced monkeys (later found out they were Vervet Monkeys), several deer, wildebeests, warthogs, baboons and some birds. They didn't actually *join* us for lunch but were so close we didn't need binoculars. All this and we hadn't even gone out on safari. The lake also contains a resident family of hippos and several crocodiles.

After lunch we were shown to our cabin which had a wonderful view over the lake. We were very close to the hippos and could hear them grunting and splashing about in the water. We watched a Pied Kingfisher diving for fish from a small tree right outside our window. We could also see Masked Weavers nesting in the tree in the middle of the lake, egrets, and a hawk.

We had to be back up at the dining area at 4 pm for tea and cakes before heading out on our first safari. There was only Una and I and a couple from Holland setting out in our jeep, which meant we could get good views. We had only just left the lodge when we stopped at the side of the road. At first, I thought we had stopped to look at a disused termite mound and I was taking some photos. Then I realised there was something much more exciting hidden in the grass – a Leopard. She was just lying dozing in the grass at the edge of the road and we were only about two feet away from her. We were told that, as long as we stayed sitting in the jeep and didn't make too much noise, we didn't bother the animals at all. In fact, it was as if we didn't really exist. It was an exciting start to our first safari. It would be impossible to remember everything that happened on our trips out in the bush but on that first night we saw four of the Big Five – elephants, the leopard, rhinos, and lions. We also saw four cheetahs, zebra, hippos, and several different types of deer. The only one of the Big Five we didn't see was the buffalo.

As for the birds – because we were only a small group and the other couple had already seen the animals, we were lucky to be able to spend more time than usual looking for birds. Since starting birdwatching Una had always wanted to see rollers, bee-eaters and hoopoes. She hoped that we would at least catch a glimpse of bee-eaters and rollers in South Africa. She got her wish when we spotted a Lilac-Breasted Roller – lilac breast, blue belly and brown wings – sitting on a branch hanging over our path. By the end of our trip to Chitwa Chitwa we had seen so many rollers (both Lilac-Breasted and European) that we were saying 'Oh look there is another roller!' However, as we continued our drive another bird flew past us and Una looked at me for confirmation – was it really a Hoopoe? It was and this was one of the birds she hadn't even thought about seeing on this trip. She thought we would have to go birdwatching in Europe to spot Hoopoes. I don't think she stopped smiling for over an hour. On our way back to the lodge we were behind another jeep and they had to break up a herd of elephants that were on the road. The elephants were not too happy about it and by the time we came to pass them one of the larger ones decided to make a run at us. Patrick just revved the engine and kept moving quickly and the elephant just trumpeted at us – telling us to clear off. What an amazing evening.

We were back at the lodge at 7:30 and just had time to get tidied up for dinner. When it was dark, we were not allowed to wander around on our own and had to be escorted to and from the dining area. One of the rangers had been attacked by a buffalo near our chalet some months previously. Patrick, our guide was knocking on the door at 8 pm and escorted us and the Dutch couple (who had the cabin next to ours) up to the bar. Another group of people had arrived while we were out on safari so there was quite a large group of us for dinner. Dinner was served in an enclosed area open to the stars and lit by candles and a log fire. Una and I were very tired as we hadn't slept for over 24 hours so we got Patrick to escort us back to our cabin immediately after dinner. There were lots of little bugs around the room but as we had mosquito nets around our beds we didn't worry too much. We had to be up again at 4:45 am to head out on the next safari.

The morning safari was uneventful. We saw the four cheetahs again and this time they were having fun chasing one another and clawing at the trees. There was also a herd of elephants breaking down trees. We even managed to catch a glimpse of the breeding herd of Buffalo but they were some distance off. We had now seen all of the Big Five. We asked Patrick about giraffe as we hadn't seen any so far and he said that they were rare here because they didn't like the lions and would move on to a place where the lions weren't about.

We were back at the lodge about 9 am for breakfast. After breakfast, we had the option of going on a walking safari but decided to leave it until the next day as we were still tired from all the travelling. We were going to have a rest but there was so much happening outside our window it was hard to rest. There was a small bush beside our patio and it was covered in large black and yellow beetles. Suddenly a bird arrived in our tree and then another and another – four in all. They were crimson in colour with a long thin tail. They began to swoop down to the bush and pick off the beetles, fly back to the tree, bash the beetle on a branch until the wings fell off and then eat it. We checked our bird books and discovered they were Carmine Bee-eaters. We saw them every day we were at the lodge.

We also saw a Red-billed Wood Hoopoe – a beautiful large black bird with a long red bill. The weavers kept flying back and forth and a Crested Barbet came to sit in the tree too. Then a wonderful bird with a very long tail landed in our tree. It turned out to be a Paradise Flycatcher – mainly rusty brown with a blue-grey head and under parts. At 4 pm it was time once again to head off for the next safari. We caught up with the breeding herd of buffalo which were spread out all over the dirt airstrip we had landed on the day before. Apparently, this is a regular occurrence and sometimes the planes have to do several flypasts in order to get the buffalo to move so they can land. Several of the buffalo had Red-billed Oxpeckers on their backs. These are small birds which help to keep the buffalo parasite free. However, they are also partial to a drop of blood and can often be seen pecking at wounds which the animals have picked up. We saw one big male with quite a large wound on his back which the Oxpeckers seemed to be keeping open. I am sure they must get infected sometimes but the buffalo don't seem to mind.

However, there was still more to come. Patrick got a call on the radio which I couldn't quite make out and he headed off to look for something. We stopped at a large tree full of vultures. They are amazing birds and so big. Then Patrick began driving over the bushes and small trees in front of us and we all wondered what he was doing. We headed down a small incline and there in front of us was a large male lion happily munching away on a buffalo. Every now and again he would tear off some of the flesh and as he did so the buffalo's head would move and the large white teeth would shine out at us. Una managed to get some good shots of it on the video. We had only ever seen this kind of thing on nature shows and never dreamt of actually seeing it in real life.

We had so many photos now that after dinner I decided to download them on to my computer in case anything should go wrong with the camera. A total of 250 so far and we had only reached day three.

The most exciting things the next morning were catching a glimpse of a puff adder as it glided past in the grass beside the jeep and a sparrow hawk flying past with a very large snake caught between its talons. We also saw a pride of lions – two females and six cubs lying sunbathing on a dry riverbed. They didn't even move when we drove up behind them.

After breakfast we went on the walking safari which took about an hour. It was very interesting and our guide (who had a large rifle with him) told us all about the different plants and the ways in which his people would use them for medicines. There was even one tree with very large thorns which could be used for sewing needles. We saw a Snake Eagle and a large flock of the Red-billed Wood Hoopoes. After lunch, we spent the afternoon in our room watching the Carmine Bee-Eaters and Woodland and Pied Kingfishers swoop down from our tree in search of food. Then it was time to head out on our next safari.

For the rest of the group in our jeep that night was their last safari as they were all leaving the next day and I think they got a very good evening. We had wonderful views of a leopard which came down to a waterhole for a drink. We followed him for some time before leaving him in peace. Then we got a beautiful photo of another leopard relaxing up a tree.

That evening the lake on which our lodge sits became the bathing ground for the breeding herd of elephants. Our jeep was close by so we got back in time to watch them. They all seemed to be enjoying the water but the little ones were having great fun diving into the lake and jumping on top of one another. Eventually the herd decided to move on and the adults headed off but the young ones were reluctant to leave. Eventually they walked out of the water only for one of them to turn round and run back in again followed by all the other little ones – typical children! This happened a couple of times before they eventually all moved off. One other exciting thing was to see giraffe. Patrick had said it was unlikely we would see them because of the lions but we got up very close to three of them browsing on the trees – their long necks meant they could reach where none of the other animals could. Considering their size, they are such graceful animals. We could have watched them all evening, but dinner was waiting. Una was joined by a large black beetle. He was sitting on her plate when we got to the table. Maybe he was hoping for some dinner but Una wasn't going to share. We chased him away.

Next morning only Una and I were out on safari as all the rest of our group were leaving. This was great as we spent most of the morning looking for birds. We saw 41 different species of birds and we were just a few feet away from a pair of Yellow-billed Hornbills which were sitting on a tree stump doing their mating display – what a wonderful sight. But best of all we saw one of our favourite birds, a Paradise Whydah, joint No 1 with the Superb Blue Wren, Australia. We both fell in love with it as soon as we saw it. Una said that if she had seen nothing else during the whole trip this bird alone would have made the trip worthwhile. It has very long flowing black tail feathers, a black back and head and a creamy brown chest and neck. We loved watching the 'Life of Birds' presented by David Attenborough and he often talked about Birds of Paradise. They are such beautiful birds and we thought how lucky he was to have seen them. Well now we had seen a Bird of Paradise too and we agreed that it was one of the most beautiful birds we had ever seen.

Besides all this excitement we were also caught up in an elephant roadblock. The track we were on was blocked by a large male elephant. Towering above our tiny jeep, his trunk swinging slowly from side to side he looked like he was spoiling for a fight. He stared at us as if to say: 'Try to pass. Make my day'! Patrick eventually decided to make a small detour through the bushes and get round him that way. The elephant didn't like it much but eventually decided it wasn't worth chasing us.

The chef was serving lunch that day and he seemed very concerned when Una told him there was a big problem with his food. However, he began to laugh when she said it was too nice. We spent the day in our room because this was our last day and we want to see as much of the Carmine Bee-eaters and Kingfishers as possible. They didn't let us down.

Our final evening safari and we were still managing to see new things. We saw a Leopard Tortoise. This is one of what is called the Little Five. There are four other small animals with the name of one of the big five. We also saw the female Leopard relaxing in the long grass. Patrick said her cub was close by but he couldn't spot it at first. Then our tracker spotted it hiding in a large bush. It was feasting on a Steenbuck deer and thoroughly enjoying it. We even came across a pride of lions, one male and six females, and we followed them as they went off to hunt. We didn't manage to see the hunt as it was now dark and we had to head back to the lodge. That night Una had one final surprise. There was a large spider, about the size of our house spiders, sitting on the mosquito net which was around her bed. She hates spiders and always dreaded seeing one in the room. Spider Man came to the rescue. At home, I would catch the spider and put it outside. I didn't know if this one was dangerous. I didn't take a chance.

Our final safari was just as good as the first one we had ventured out on four days ago. We saw something totally unexpected. We came across the breeding herd of buffalo. The herd seemed to be quite agitated about something. Then some of the buffalo began running and at first, we couldn't see why. Two male lions were in the midst of the Buffalo herd but instead of the lions trying to catch a buffalo, the buffalo were chasing the lions. It was so funny to see the two big males literally running for their lives. We later saw the two lions having a well-deserved rest after their lucky escape.

The mother Leopard and her cub were on the side of the path leading down from our lodge and the cub was very playful. He kept running away and then bounding back to jump on his mother. She didn't seem to mind too much.

Back to the lodge at 9 am for breakfast and then we had to pack. Luckily, we still had the use of our cabin as no other guests were due to arrive until the next day. I finished typing up our list while Una packed the bags. While eating lunch we were watching buffalo, elephant, giraffe, zebra, wildebeest, deer, hippos, and a fish eagle all just on the other side of the lake from where we sat. I spotted the resident crocodile gliding past with an Impala antelope clasped tightly in his jaws for his lunch. We paid our bill and Patrick was waiting to take us to the air strip. However, our path from the lodge was once again blocked by three large elephants. This time we didn't argue with them but took another route.

And so ended our first adventure. The seven-seater was waiting for us when we arrived. Two other people arrived soon after and we said goodbye to Patrick. We had one more pick-up on the way at another lodge and then it was back to Johannesburg for an overnight stay before heading to Victoria Falls.

We arrived in Johannesburg at 4:30 pm and once again we were surrounded by porters. However, we were glad of them as we needed to pick up Una's big case. One porter took us to the left luggage place and then asked us if we needed a taxi. I explained that we had to phone the B&B and get someone to collect us. The porter very kindly took out his mobile phone and made the call for us. The van arrived about 20 minutes later and we had a short drive to the B&B. The B&B was very comfortable and we decided not to go out for anything to eat, but just to have a sandwich in the room. Tea and coffee were provided. Una was happy to be back in a place where there was no need for mosquito nets.

Victoria Falls

We had a relaxing evening watching a bit of TV. We went to bed early as we were both very tired after all our early mornings. We were back to the airport at 9 am and checked-in for Victoria Falls. We had a look around the shops, bought presents and postcards and then decided to have a drink. We were just sitting down when we spotted the Dutch couple who had been in Chitwa Chitwa with us. They were also going to Victoria Falls on the same flight and not only that but they were sitting in the same row as us but on the opposite side of the plane. They were staying in Victoria Falls until Saturday and were getting the same flight back to Johannesburg as us. The flight was 1 hour 15 minutes and was very smooth.

There was someone waiting to take us to our accommodation which turned out to be a safari lodge set in a forest. It was very hot (over 30^{0}C) and we were glad to get into the jeep. We drove through Livingstone town but didn't get a chance to visit it. The lodge was about 15 minutes outside the town and about 10 minutes from the falls themselves. The lady who had picked us up asked us what we would like to do while we were there. Of course, we wanted to visit the falls and Livingstone Island and maybe do a safari of some kind. She was new to the lodge and began to tell us about some trips she had done and would recommend. She told us how she was an avid birdwatcher (lucky for us) and all about a day trip she had just done to Botswana. Needless to say, we opted to do that on the Friday. Two big dogs came to meet us when we arrived (Gin and Tonic) and they reminded us of Lucy. We had contacted the kennels and they told us she was fine. Our room was like a cabin made from mud with a grass roof. It was nice but there was a large gap under the door which led out onto the balcony. Una was concerned that anything could have gotten in, never mind spiders! However, we didn't see any bugs in the room the whole time we were there.

After lunch we went on a tour of the falls with a guide called Victor. He was very knowledgeable and told us all about the falls. As well as working for the lodge he was also a local football coach. On our way down to the falls we bumped into the Dutch couple again. They were staying in the hotel at the falls and were just on their way down to get some photos. The falls were absolutely amazing. We walked over a bridge from which you could see a beautiful rainbow. Our guide told us that, sometimes, you can see a full circle rainbow but we didn't manage to see one. We were soaked to the skin but it was really worth it. However, we both agreed that Iguasu Falls (in South America) was still No. 1 for us.

We stopped to look at the curio stalls on the way back but the men kept trying to get us to buy everything. They would put items into our hands and tell us they would make us a very good price. In the end we bought a few things (mostly carved animals) and paid far too much money for them. But at least the money was going to the local village. We went back to the room to change and decided to go and sit by the pool for a while before dinner. Little Bee-eaters were swooping in the trees just in front of us and so were Blue Waxbills. Even a Paradise Whydah came and sat in a tree, just feet away. An Emerald Spotted Dove came to have a drink from the pool.

It was early to bed as we had to be up at 6 am the next morning for the trip to Botswana.

Botswana – The Zambezi River

We had a long drive and two passport checks to get to Botswana. Then a short boat ride over the Zambezi River and a jeep to the lodge where we were to get the boat for the Chobe River cruise. There were birds everywhere. We saw a beautiful bright red bird on the Zambezi which turned out to be a Red Bishop, loads of Yellow-Billed Kites, and we passed a waste dump on the way to the lodge which was surrounded by Marabou Storks looking for an easy meal.

We were the only two people going out in our boat which meant we could look at what we liked. There was tea/coffee, cake and biscuits waiting for us on board. We set off across the Chobe as there was a Fish Eagle in a tree on the other side. He was a beautiful bird with a white tail and head and brown body. Even though we had seen Fish Eagles in Chitwa Chitwa this was even more impressive as we were right underneath it.

The boat now decided to stop working and the guide called to one of the other boats to come and help. He couldn't sort out the problem so phoned back to the lodge to get another boat. The new boat wasn't as nice as the first one and it had a large cockroach on board. I tried to catch it but it was so fast it kept getting away. We eventually put a cup over it and threw it overboard only for another one to join us a short time later. There was so much to see that Una soon forgot about it. We spent about three hours sailing up and down the Chobe and we really enjoyed it. We saw European Roller, Lilac-Breasted Roller, Blue-Cheeked Bee-Eater, Little Bee-Eater, White-Fronted Bee-Eater, Pied Kingfishers, Fish Eagles, Large Egret, Cattle Egret, Black Egret, Squacco Heron and African Darter to name a few of the birds. We also saw Vervet monkeys, two elephants crossing the river and a mother hippo with her very small baby. She didn't like it when one of the boats got too close and made a dive at it. The guide had to reverse quickly to get out of the way. Not far from the hippo a crocodile was basking in the sun on the bank of the river. We got up very close to him and Una suggested that I stand up on the front of the boat to get a good photograph. The crocodile didn't like this and began spitting.

It was a most enjoyable morning and we were back at the lodge at 12:30 for lunch. After lunch we had time to look in the shop where we bought a new bird book, some post cards and a shirt for Una. Then it was time for the afternoon safari drive in Chobe National Park. Again, we saw all the usual suspects – elephants, hippos and baboons. We stopped at a small waterhole to watch a family of warthogs having fun playing in the mud. I noticed a white bird in the waterhole. I thought it was an egret until it lifted its head and I could see its bill. It was a Spoonbill. Like the Bee-eaters and Rollers this was one of the birds which was high on Una's "must see" list. She wasn't expecting to see one on this trip. Standing beside the Spoonbill was a Hammercob, a strange-looking bird with a very big bill. So many "must see" birds in one short week. What an achievement. Una said that if she never saw another bird, she would still be very happy. On the way back to Stanley Lodge we ran into a thunderstorm and some rain but it didn't matter as we were on the bus. All in all, another amazing day.

We decided to go to Livingstone Island next morning – a small island in the middle of the falls. We had a quick breakfast and headed off in the bus. We got a boat at the Livingstone hotel (a beautiful hotel right on the falls – Una said that next time we will stay there) and had a very short trip over to the island. Here we were met by a guide and taken for a tour of the island. Once again, we were soaked as we were able to get right up to the edge of the falls and look over.

In the summertime, when there is less water going over the falls there is a pool you can walk to from the island and swim in. When we were there, it was closed. We didn't realise that breakfast was served on the island. This was special – what a setting, listening to the water pouring over the falls.

On the way back to the boat we stopped to have a look at the "Loo with a view" and what a view looking right out over the falls. It must be amazing when the river is in full slate.

We were back at the lodge at 9:15 am and had plenty of time to pack and have a quick look in the shop. Una bought a couple of bags, one of which she said she would use for work. Then it was time to head to the airport. Once again, the Dutch couple were checking in just in front of us and we sat chatting to them while waiting for the plane which was quite late. Again, they were sitting in the same row as us on the way back to Johannesburg. They were heading home that night so we wouldn't see them again.

Una spent the flight writing post cards and we landed in Johannesburg at 3 pm. This time we didn't need a porter and phoned the B&B to get the bus to pick us up. We got the driver to drop us off at a shopping centre so we could have a look around and he took the luggage back to our room. We had a coffee first and then wandered round the centre. However, most of the shops were closing as it was almost 5 pm. Una was looking at shoes for me but nothing suited so we just bought some water and crisps and phoned for the bus to come back for us. We stood outside the centre waiting for the bus and there was a thunderstorm. It didn't rain and it was fascinating to watch the lightning flash across the sky. The thunder rumbled on most of the evening.

The following morning, we were on our way to Cape Town.

Cape Town

We could hardly believe that 10 days of the trip had passed so quickly. We had an early flight and Una finished writing postcards. The flight took 2 hours 15 minutes and we landed in Cape Town at 9:45 am. Belinda, one of my judicial friends, picked us up and drove us back to her house where we were staying until Wednesday. William (her husband) was there when we arrived and we had coffee sitting out on the patio. They have a beautiful house with a lovely garden and swimming pool. William showed me a car he had (which belonged to his mother) and told us we could use it if we wanted to. We decided to head into the Waterfront as Una wanted to get some presents for our sister, Sheila, and Una's friend Roisin, so Belinda said she would take us to show us the way.

I took Una to the jewellers where I always buy presents when in Cape Town and we picked out necklaces for the two girls. We decided to have coffee at a restaurant nearby and sat outside in glorious sunshine with beautiful views of Table Mountain. This would have been a great day to be on top of the mountain. We spent the rest of the afternoon wandering around the Waterfront while I pointed out all the things I liked about the area.

By now Una had fallen in love with Cape Town. We booked a trip to Robben Island and then rang Belinda to come and pick us up again.

That evening we were going to have dinner with a friend of Belinda's who was staying in an apartment with the most amazing views over the bay. After a wonderful dinner, served on the balcony, we sat chatting and watching the sunset.

The following morning, we decided to head for Table Mountain. You could see the mountain from Belinda's garden and it was quite cloud-laden but by 9 am the cloud seemed to be lifting a bit so we took a chance. Belinda moved her car to allow me to get the 43-year-old Volvo out but it wouldn't start. When William was showing me how to use the car the day before he had left the lights on and the battery was dead. Belinda said we could use her car instead as she was working from home. We drove up to the cable car station and had a look in the shop. We bought loads of soft toys for all the kids. The lady in the shop asked Una if she was buying a zoo. The cloud had lifted slightly now so we bought our tickets and headed up in the cable car. Before we got to the top the cloud had come down and it was raining again. It was cold and miserable so we went and had a coffee and headed back down. At least we still had four more days to try again.

Ever since coming to Cape Town six years before I had talked about my favourite walk along the sea front where I found a Blacksmith Plover's nest and saw Black Oystercatchers (which are on the endangered list). I told Una I would love her to see them too. So, we headed for the seafront and spent the next two hours walking. We spotted loads of cormorants and the Blacksmith Plover sitting tight on her nest in the usual spot but no sign of the Oystercatchers. We decided to turn back. We spotted an Egret and stopped to look at that. The Egret moved a few steps and I spotted an Oystercatcher right behind it. There was just one but Una got a good look at it. We were both happy that she had seen what we came looking for.

We headed back into the Waterfront as we were going to Robben Island and we wanted to call at the jewellers to ask them to make a golf 'green repairer'[34] for our nephew Ciaran who would be making his First Holy Communion in May. Then we headed to the ferry. The trip out to the island took an hour. We sat out on deck although it was quite cold and windy and the sea was a bit rough.

When we arrived on the island there was very little organisation. We just followed the people in front and we were all piled onto buses which eventually took us for a short tour of the island. They brought us to the quarry where Nelson Mandela and many other prisoners spent years quarrying limestone. We weren't allowed to get off the bus at any stage so we didn't really see much or get any good photographs.

After the bus tour we were dropped off at the prison and one of the ex-prisoners took us on a tour. It was very interesting. We started off in one of the larger prison cells which housed around twenty prisoners at a time and we were told about the differences in the food rations between the black prisoners and the coloured prisoners and also how the black prisoners weren't allowed to wear long sleeved shirts winter or summer. Then we were taken to see Mandela's cell but because there was such a crowd of us, we didn't really get a good look around. We would have to content ourselves by reading his book 'Long Walk to Freedom' again. At least it would be more meaningful now.

---

[34] The putting green on a golf course requires good maintenance, and that is why a green repair tool is a key part of any golfer's or greenkeeper's equipment.

Then we were left at the harbour and it was time to get back on the boat. We didn't even have time to look in the shop for souvenirs. We decided to sit inside for the return trip as we had no jackets with us and it would be cold on deck. Everyone had the same idea and there wasn't much room. The journey back was rough and some of the groups weren't looking too happy. For once the bumpy journey didn't bother either of us. Nonetheless we were glad when we got back on dry land.

We headed back to Belinda's. William had prepared a beautiful dinner. Belinda's brother and some of their friends joined us. It was a pleasant evening and the food was excellent again. William is a great cook.

On Tuesday we decided to head for Cape Point. Again, we were able to use Belinda's car. Our first stop was at Boulders Beach to see the Cape Penguins. Visitors were restricted to a walkway overlooking the beach. There were penguins all over the beach – no need for binoculars here. It was quite windy and my hat blew away. Luckily there were some workers close by and they were able to get it for me. We spotted a pair of Black Oystercatchers with two chicks down amongst the penguins. On the way back Una spotted another beach where people were allowed. The penguins didn't seem to mind as we wandered around taking photographs.

We had coffee and bought some presents. Una collects spoons. She got a glass spoon with a penguin on it which is unique to Boulders Beach and then we headed for the Cape of Good Hope.

In the car park at the Cape there was a baboon intent on causing trouble. An elderly couple were heading up towards the entrance and the baboon tried to grab the bag which the lady was carrying. Luckily, she managed to keep hold of her bag and he ran away. There was a cable car up to the lighthouse but we decided to walk even though it was quite hot. We wanted to see as much as possible. However, the animals and birds must have thought it a bit hot too as there weren't many about. When we got to the viewpoints we saw a couple of lizards, a small mouse and even a glimpse of a snake as it slithered off the edge of the path in front of us.

We had a look in the shops and bought more presents and some crisps and water and decided to head for the Cape of Good Hope and have a picnic. The big baboon was still in the car park but must have managed to steal some crisps and Fanta. He was sitting on the roof of a car where he had spread all the crisps out and was dribbling the Fanta over the top of them and licking it off. The car must have been a mess when the owner got back.

It was a short drive to the Cape of Good Hope and we saw a female Ostrich on the way. We took some photos of the sign and sat on the beach eating our picnic and watching the terns and cormorants. It was a great experience to be there.

It was time to head back to Cape Town but we had a quick stop at an Ostrich farm on the way to get some photos of the males. There was a baboon here too trying to steal the ostrich food but they were having none of it.

Belinda had to go to a meeting that evening so we went out for a Greek meal with her and then spent the evening watching TV with William.

Wednesday morning dawned bright and cloud-free and, as we were leaving Belinda's that day, we decided to make one last use of her car and head for Table Mountain. We set off about 9 am and it was already quite hot. There was quite a queue this time but it didn't take too long to get to the ticket desk and up the stairs for the cable car. We had to wait about five minutes for the car to come down.

On the way up the cable car turns round 360^0 so everyone gets to see the views over the bay. We stopped at the cafe for a drink and then headed for a walk along the top of the mountain. There weren't many birds or animals about but we did see some sugarbirds and a couple of lizards and the most beautiful cricket which was bright red in flight. As it was such a hot morning, we didn't walk too far but spent about an hour looking at the amazing views. Una said she could see why I liked this place so much.

After our walk we headed back to the Waterfront as we wanted to check on the jewellery. Everything was going fine so we decided to head back to Belinda's and pack our cases. On our way back to the car Una spotted some dolphins swimming about in the harbour. We stood watching them for a while. I told her I had never seen dolphins in Cape Town before.

Zenobia, another of my South African friends, had invited us to stay with her for a few days. She is a lawyer who specialises in Family Law. Belinda dropped us off at 3 pm. Zenobia's two daughters were just getting home from school. They are two lovely girls and both are very active. They spent the afternoon running around the garden and doing cartwheels. Zenobia and Chris (her husband) were taking us out for a meal together with some colleagues of

Zenobia's. We went to a lovely restaurant at Victoria Wharf with spectacular views of Table Mountain. We had a very enjoyable evening.

Zenobia goes to work very early in the morning. Before leaving, she said that she would phone us to tell us what time we could pick up her car (which was in for a service) so we spent the morning relaxing in the house. They have a beautiful home nestling in the hills just below the 'Lion's Head'. We sat in the garden watching the pigeons and doves roosting on the roofs opposite.

Zenobia sent one of her colleagues to pick us up about 9:30 and drop us off at their office. From there we walked to the Waterfront and got a bit lost on the way. We spent the rest of the morning wandering around the shops as Una still had loads of presents to buy. She got some beautiful carved horn and shells and even a lion's tooth for herself (which is now attached to her mobile). After lunch, we took a taxi up to Green Market Square – a large square which was filled with stalls. It was interesting wandering around but we had bought everything we needed and more. We had to go back to Zenobia's office to get her car keys and then a car from the garage came to pick us up. He had to pick up another person and by the time we eventually got through the rush-hour traffic we were totally lost. We were some time in the garage as no one seemed to know what was happening. Eventually we got the car and headed off to try and find our way back to the house. It wasn't as bad as we feared and we found our way on to Strand Street (where the office is) quite easily. It was a straight run from there. Zenobia and Chris didn't get home until after 7 pm and they had left the house around 6 am! They work very long hours. Zenobia made dinner and we spent the evening chatting. She had taken home from the office a large suitcase full of files which she needed to read for the next day so at 10 pm we said we would go to bed and let her get on with her work.

Our last full day of the holiday and we had arranged to meet up with another couple of my friends for lunch. They were going to pick us up at the house but we phoned to say we would meet them at the Waterfront. We headed down to the sea front and spent the next two hours walking along looking for Black Oystercatchers again and they didn't let us down. We saw about four of them. It was hard to believe we were going home the next day but seeing all the people walking their dogs made us miss Lucy. We met up with Jackie at 12:30 pm and she drove us out to Hout Bay for a lovely lunch in a fish restaurant set right on the beach. Daksha was waiting for us and we had a pleasant afternoon eating wonderful food and chatting. We had a walk around the harbour before Daksha left us back at the Waterfront.

We called at the jewellers to see if the items we had ordered were ready. One of the staff was on his way back from the engravers and arrived just after we did. The 'green repairer' we ordered for Ciaran looked really good with his name, Cape Town and the year on it. The silver spoon Una got was lovely too and the emerald ring she ordered (which was hand made since Monday) was beautiful.

We spent a pleasant evening with Chris, Zenobia and the girls and watched a procession of people climbing the Lions Head to see the sunset. Later on, we saw the procession of torch lights as they came back down again.

Bia (the eldest girl) had a school sports event next morning and the girls had to be at the school for 6:15 am. A very early start for young children but I suppose the staff were thinking of the heat later in the day. We had to go to the Waterfront to sort out our tax refund and, as we were a bit early, we had a coffee sitting by the harbour. There was a big seal sunning himself on one of the boat walkways and a man who was washing his boat decided to chase him. He turned his hose on him until the seal dived off into the water. The seal just swam around until the man went away and promptly climbed back onto the walkway to continue sunbathing.

We got our tax refund sorted out and then had to go to the tax desk in the airport to get the money back. We spent the rest of the morning packing our cases and making sure all the breakables were in the hand luggage.

Then Chris, Zenobia and the girls took us for lunch to a lovely restaurant on the sea front at Camps Bay. It was really beautiful and I think the girls enjoyed it but they were both very tired after their very early start. We took a few photos on the beach before Chris dropped Zenobia and the girls off home so they could have a rest while he took us to the airport.

Luckily, we had plenty of time as there were a lot of road works on the way. Chris dropped us off and we first had to go to International Departures to get our tax refund. They asked to see the goods we had bought so Una showed them her ring and they didn't ask to see anything else. Una had most of the things in her hand luggage just in case. They gave me a cheque which I had to cash in Johannesburg airport. Three different places just to get some tax back. Then we had to check in at domestic departures for our flight to Johannesburg. We had an uneventful 1½-hour flight and as soon as we landed, we went looking for the place to cash the cheque. We found the tax refund place and there was a long queue. Una suggested I try the exchange place next door. They cashed the cheque for me and charged me quite a fee to do so. Then Una went to the duty free and got some cigarettes for her mum before we went down to the South African Airways lounge. We checked if it was possible to upgrade but the flight was full. We took off for London about 10 pm and had a long 10½ hours ahead of us. We flew through a thunderstorm and Una watched the lightning out her window. The flight was quite bumpy for about four hours but wasn't too bad. Una watched a couple of movies and I read another book. We landed at 7:30 am and made our way through security. We were only allowed one piece of hand luggage going through so Una had to take her two small bags and stuff them into the one she was carrying with the presents in it. Then when we were going through the metal check, we had to take our shoes off and put them through the X-ray machine. We made it to the BMI lounge and were glad to sit down and have a cup of coffee.

Our next flight was due to take off about 10:40 am so we made our way to the gate at 10:10 am but there was no sign of movement. We bumped into John and Janet (friends of mine from the court) and sat talking to them. The plane eventually arrived and it turned out that they had been fog bound in Belfast and there might be problems landing on the way back. The captain assured us he was carrying enough fuel to allow us to circle over Belfast for an hour if necessary. However, it wasn't necessary. As we had been so late taking off the fog had lifted just enough to let us land. We picked up our cases and headed for a taxi. It was freezing, about 2°C, and we had just come from +30^{0}C. We queued for about 15 minutes and were glad to get into the taxi. The house was freezing when we got in so I put the heating on full blast. I also turned on a small electric heater and the gas cooker in the kitchen and Una lit the gas fire and electric heater in the living room. We were both very tired but decided not to go to bed as we probably wouldn't sleep later. Una unpacked her case and sorted out all the presents. We managed to keep going until about 8 pm and then gave up and went to bed.

I was working the next day in court but we were out at Ballyharvey to pick Lucy up at 8:30 am. She was delighted to see us but we got such a telling off all the way home. She just kept howling at us as if to say, 'Why did you go away and leave me?' However as soon as we got home all was forgiven. A perfect end to a wonderful holiday.

Sweden
1996

I was in Stockholm, Sweden, in August 1996 to attend the first World Congress against the Commercial Sexual Exploitation of Children. Stockholm is a beautiful city – "Beauty on water". 30% of the city area is made up of waterways and another 30% is made up of parks and green spaces. But I had no time to explore it. All my attention was focussed on the World Congress. The best I could do was to walk from my hotel to the conference centre and back again. I saw enough of the city to make me want to come back and spend some time there. It didn't happen.

When I think of Stockholm I think of the conference and of the horrific stories I heard of the commercial sexual exploitation of children. You will find a report of the Stockholm Congress (and of the second World Congress in Yokohama (Japan) in 2001) in the final chapter – "A World Apart").

Apart from those stories one other incident springs to mind. One day I was delayed getting back from lunch. I didn't want to interrupt proceedings so I stood at the back of the lecture theatre waiting for a break before taking my seat. Then I noticed someone else had slipped in and was standing beside me. I looked round and realised it was Roger Moore. I reached out my hand and said: "Hello Simon". He smiled as he shook my hand. I said: "I'm a Saint fan, not a Bond fan". We chatted quietly for a few minutes and he told he was there in his role as a UNICEF Ambassador. My son (a big Bond fan) was sorry he hadn't been with me at that conference.

Switzerland

A Country with Many Names

Suisse – Schweiz – Svizzera – Svizra – Helvetia

In my report on Myanmar/Burma (p145) I asked: How come this country has two names? Now we find that Switzerland has six names! Four of these we can put down to its linguistic diversity. Switzerland has four national languages: German, French, Italian and Romansh, each providing a different name: Schweiz (German); Suisse (French); Svizzera (Italian); and Svizra (Romansh).

All four names appear on Swiss banknotes. But there isn't room on coins or stamps for four names. In some situations, *Latin* is used, particularly as a single language to denote the country. Confoederatio Helvetica is the nation's full Latin name. That name is derived from the Celtic Helvetii people who first entered the area around 100 BC. In 61 BC the Helvetii, under the leadership of Orgetorix, tried to take over part of Gaul, as the lands were much more fertile. They ran into a formidable opponent in Julius Caesar who promptly chased them back home. Helvetia was the Roman name for the region that is now western Switzerland. This is the name that appears on coins and stamps. The inscription on the front of the Federal Parliament in Bern reads "Curia Confoederationis Helveticae". The data code for Switzerland, CH, is derived from the Latin Confoederatio Helvetica. And vehicle number plates bear the letters "CH" to indicate that they are registered in Switzerland.

The country is known to the English-speaking world as Switzerland. Where did this name come from? No one is quite sure. There has been lots of research and discussion but no definitive answer. The German name Schweiz is believed to be derived from the name Schwyz, one of the first three cantons that got together in 1291 to defend themselves against the Austrian Habsburg Empire. Since the French- and Italian-speaking areas only became part of Switzerland in the early 19th century, I would be inclined to accept that the name evolved from Schwyz.

My first visit to Switzerland was in 1969. I was touring Europe by caravan with my wife, Bernie. Like Caesar before us we were just passing through and decided to stop in Geneva overnight. To Caesar we owe the first written mention of Geneva – in De Bello Gallico he briefly refers to "Genaua". He is said to have camped right in the centre, roughly where the Placette department store is now. We found a caravan site not too far away.

My wife was diagnosed with MS on our return home and it was to be 20 years before I was in Switzerland again. In December 1989, I was in London for a conference. I decided to fly to Geneva and spend the weekend with my son, Liam, who was studying at the university there. He took me on a guided tour of the city on board the Tram Douze (The No.12 Tram). We visited the University and had long walks by the lake. On the Sunday afternoon we had a very pleasant walk around the small lakeside town of Nyon, about 10 minutes away by train in the direction of Lausanne. Nyon also goes back to Roman times, and there is a Roman museum there.

That was the weekend of the Course de l'Escalade, the annual fun run which is held a week before the Fete de l'Escalade proper. "Escalade" means climbing and refers to the Duke of Savoie's attempt to sack the city of Geneva in 1602. The assault was successfully beaten back by the locals. La Mère Royaume, a semi-legendary lady, turned out to be the unlikely heroine. It is said that she had been preparing a huge pot of soup in the middle of the night, and poured the hot soup over the invaders trying to scale the fortifications!

Geneva lies at the southern tip of Lac Léman (Lake Geneva). Surrounded by the Alps and Jura mountains, the city has dramatic views of the Mont Blanc. With the European headquarters of the United Nations and the Red Cross, Geneva is a global hub for diplomacy and banking.

I joined the Executive Committee of the IAYFJM in 1994. Jean Zermatten, a Swiss judge and one of the most dynamic members of the Association had founded *The International Institute for the Rights of the Child / Institut international des Droits de l'Enfant (IDE)* in his hometown of Sion. IDE is a private Swiss Foundation. Its role is to train professionals who work with and for children, so that children's rights may be better known. To that end it organises courses and seminars throughout the year and one major international conference annually – generally in October. Sion is in the Canton of Valais, about 90 minutes by train from Geneva. Members of our Executive Committee generally attended the annual conference and facilities were made available for us so that we could hold a meeting afterwards.

I attended the annual conference in Sion every year from 1994 to 2004. Generally, I was presenting a paper, or chairing a workshop – sometimes both! Liam was living and working in Geneva so I could count on him to take a day or two off and assist with interpretation – on a completely voluntary basis. On one occasion he stayed for the entire week. I remember we worked to 3 am to perfect the final report.

We generally stayed at the Hotel Europa, just a few minutes' walk from the train station. We were collected each morning and taken by bus to IDE and left back to the hotel in the evening. We had the opportunity to socialise during conference breaks and at dinners out in the evening. Jean was an excellent host and would organise fondue or raclette evenings, sometimes in town, sometimes outside. He had a vineyard and produced his own wine. At some point during the conference, he would invite the Executive Committee to join him and his family for dinner in his own home. On one occasion he invited Liam and his partner, Lisa, to join us.

I enjoyed all the after-conference outings Jean arranged for us – Raclette and Fondue have remained two of my favourite dishes since that time. But one very special outing was to the thermal spa resort of Leukerbad. The Romans were among the first to recognise the healing properties of the hot springs. Every day 3.9 million litres of hot water gush from 65 thermal springs making Leukerbad one of the biggest spa and wellness resorts in Switzerland.

I was enjoying a swim in an indoor pool when Jean told us to follow him. We dived under a barrier and found ourselves in an outdoor pool. It was a memorable experience to be swimming in an outdoor pool lit by a bright harvest moon on a cold frosty night in October.

In 1998 I was elected Vice President of the IAYFJM. I took on responsibility for arranging passes for members attending various UN meetings relevant to the Association's work. So, I would usually be in Geneva at least twice a year, in spring and autumn. Liam was always keen to introduce me to new things such as attending the Geneva motor show or going birdwatching in the botanical gardens.

Sometimes we visited Michel Lachat, another Swiss judge and member of our Executive Committee. Michel lived near Romont, in the canton of Fribourg. He showed us around Romont and we visited the famous stained-glass museum.

One of my former students had married a Swiss national and had settled in Olten. When time permitted, I would take a trip to Olten to visit her and her family. This gave me the opportunity to see another part of Switzerland.

As we approached the end of 1999 Liam was keen that Una and I would come to Geneva and see in the new Millennium with him, Lisa and Lisa's parents. He came home for his usual Christmas visit. The three of us flew to Geneva on December 27.

Natasha, a friend of Liam's, offered Una and me the use of her apartment for the duration of our visit. Natasha's family are Russian but had been living in Geneva for some years as Natasha's dad held a senior position in the UN. The evening of our arrival they invited Liam, Una and me to their home where they had prepared a Christmas dinner for us with some Russian specialities.

Since this was Una's first time in Geneva, we spent the next day looking around the city. On December 29 we paid a visit to the historic town of Gruyere, famous for its cheese. On December 30th Lisa joined us and we headed for the ski resort of Les Diablerets, where I had booked a sleigh ride. We were disappointed to find that the resort was closed due to stormy weather. We went to Leysin instead. They had no sleigh rides with dogs so we booked a horse-drawn carriage. It was a memorable trip, very scenic, but very cold. We rounded off the evening in a beautiful, and warm, restaurant where we introduced Una to raclette. She agreed with me that we should add it to the menu when we returned home.

On New Millennium Eve we all went to Lisa's parents' place where we enjoyed a fondue chinoise[35] before heading down town. We had planned to take the bus but the buses were all packed so we decided to walk. Millennium Midnight saw us in the centre of Geneva, at a very crowded Rond-Point de Plainpalais, along with Lisa, her parents and her friend Carmen. The celebrations began once the clock struck midnight. The champagne was flowing freely and everyone was having a good time. Neither Una nor I drink alcohol and we were tired after all the travelling. Once the fireworks display ended, we decided to

---

[35] A fondue with sliced meat cooked in beef or chicken stock.

head "home". Lots of other people had the same idea and the buses were crowded again. But we managed to squeeze in.

On New Year's Day 2000, 10 of us had a meal at the Mövenpick Hotel – Una and I together with Liam, Lisa, Lisa's parents (François & Josette) and my friends, the Mueller family, who had come all the way from Olten to spend a few hours with us. We then walked around town, finding the streets positively strewn with champagne bottles, something I've never seen before or since.

We spent the rest of the weekend looking round the shops, exploring the old town and birdwatching down by the lake. We walked as far as the botanical gardens where we always see a good selection of birds.

I asked Una if she would like to see where I spent so much time when I came to Switzerland. On Monday, January 3rd, Liam, Una and I took the train to Sion. It was a cold, clear day with a light dusting of snow. After looking round the shops we took a walk up to the castle. On the way we were happy to see some Alpine Choughs. Choughs back home have a red bill and red legs. Alpine Choughs have a yellow bill and yellow legs.

As we headed back to Geneva, Una said she had enjoyed her first visit to Switzerland. It was especially memorable because it coincided with seeing in the new millennium.

My workload for the IAYFJM was considerably lighter when I stepped down as President in 2006. But my work with UNICEF and UNDP (UN Development Programme) meant that I was still in Geneva about twice a year. I continued this work up to 2012. That year I completed three out of five judicial training programmes in Turkmenistan. I was due to complete my work there in 2013, but a dispute between UNICEF and the Turkmen government led to the cancellation of the training programme. This was to be my last work-related trip to Geneva. In December 2014 I was diagnosed with Parkinson's. Around the same time, I was told my arteries were so badly blocked that I needed a heart operation. In August 2015 I was taken into hospital for a quintuple heart by-pass. I decided it was time to retire from my hectic work schedule.

Tajikistan
2005, 2008, 2009

Tajikistan is a country in Central Asia surrounded by Afghanistan to the south, Uzbekistan to the west, China to the east and Kyrgyzstan to the north. The traditional homelands of the Tajik people include present-day Tajikistan as well as parts of Afghanistan and Uzbekistan.

The area has been ruled by numerous empires and dynasties, most recently the Russian Empire and subsequently the Soviet Union. Tajikistan became a Soviet republic in 1929. When the Soviet Union disintegrated in 1991, it became an independent sovereign nation. A civil war followed, lasting from 1992 to 1997. It is now a presidential republic with an estimated population of 9.3 million.

Dushanbe, the national capital, has a population of around 800,000. I had three missions to Tajikistan and was based in Dushanbe for all three.

I undertook my first mission from September 24 to October 1, 2005. I organised and taught a judicial training programme for The Judicial Council of Tajikistan. I got on very well with the judges there and with the local UNICEF representatives. Some of the lady judges wondered why I had not remarried when my wife died. They said it was not good for me to be alone and offered to find a wife for me. I thanked them for their concern but didn't follow up on their offer. I was invited back for two further missions in 2008 and 2009. One of the judges paid me a nice compliment in 2009. She came up to me and said that I probably didn't remember her but she had attended the course I taught in 2005 and said she had learned a lot from it. She had since been promoted to the Supreme Court. When she heard that I was coming back in 2009 she immediately put her name down to attend the course as she was quite confident that she would benefit some more.

I found a certain nostalgia amongst the group for "the good old days" of the Soviet Union when everything was predictable. There may not have been a wide selection of goods in the shops but you could always get what you needed if not what you wanted. Now you can no longer be confident that you will get what you need. One lady told me that, in Soviet days, she had two weeks' holiday every year. That's no longer possible.

But my biggest surprise came when my driver told me that he was a Doctor of Medicine. He gave up his job in the local hospital to become a driver for UNICEF, because the pay was better. He had to consider his wife and children.

Tajikistan has a transition economy that is highly dependent on remittances, aluminium and cotton production. According to some estimates about 20% of the population live on less than US$1.25 per day. Migration from Tajikistan and the consequent remittances have been unprecedented in their magnitude and economic impact. It has been estimated that remittances account for 49% of GDP. Tajik migrant workers abroad, mainly in the Russian Federation, have become by far the main source of income for millions of Tajikistan's people. In 2010, remittances from Tajik labour migrants totalled an estimated $2.1 billion US dollars.

It was the view of the experts in 2010 that "The current economic situation remains fragile, largely owing to corruption, uneven economic reforms, and economic mismanagement. With foreign revenue precariously dependent upon remittances from migrant workers overseas and exports of aluminium and cotton, the economy is highly vulnerable to external shocks."

Nonetheless, my driver was optimistic about the future. Number 1 on his wish list was a trip to London and having his photograph taken beside Big Ben.

Despite my three missions to Tajikistan, I had never managed to set aside time for sightseeing. The country is known for its rugged mountains –mountains cover more than 90% of the country. Hiking and climbing are popular. The Fann Mountains, near the national capital Dushanbe, have snow-capped peaks that rise over 5,000 metres. The call of the mountains was enticing but I didn't have time to respond.

The Iskanderkulsky Nature Reserve, just 124 km from Dushanbe, is a 300 square kilometre (120 sq miles) tract of land. It takes its name from Iskanderkul, a turquoise lake formed by glaciers. The lake lies within the reserve at an altitude of 2,195 metres (7,201 ft). Claimed to be one of the most beautiful mountain lakes in the former Soviet Union, it is a popular tourist destination. But of most interest to me was the fact that over half of the reserve, comprising 177 square kilometres (68 sq. miles), has been identified by BirdLife International as an Important Bird Area because it supports significant numbers of the populations of various bird species, either as residents, or as breeding or passage migrants. These include Himalayan Snowcocks, Himalayan Ruby-Throats, Saker Falcons, Cinerous Vultures, Yellow-bellied Choughs, Hume's Larks, Sulphur-bellied Warblers, Wall-creepers, White-winged Redstarts, White-winged Snow Finches, Alpine Accentors, Rufous-streaked Accentors, Brown Accentors, Water Pipits, Fire-fronted Serins, Plain-mountain Finches, Crimson-winged Finches, Red-mantled Rose Finches and White-winged Grosbeaks.

I think that my driver was feeling sorry for me as we reached the end of my third mission and I still hadn't managed any sightseeing. He offered to take me to see Hisor Fortress, a well-preserved 16th-century fort perched on a hilltop about 27 km west of Dushanbe. Shrouded in myth and linked with many legends, the area is said to have been occupied by Cyrus the Great (600-530 BCE/BC)[36] and Caliph Ali (656 and 661 CE/AD) On the way back we stopped at the statue of Ismail Samani, an impressive bronze statue of the national 10th century hero.

The Samanids emerged as a powerful force under the leadership of Ismail Samani. They took control of Transoxiana – the ancient name used for the portion of Central Asia corresponding approximately with modern-day Uzbekistan, Tajikistan, southern Kyrgyzstan, and southwest Kazakhstan (892-907 CE/AD) and Khorasan – comprising the present territories of north-eastern Iran, parts of Afghanistan and much of Central Asia (900-907 CE/AD).

Ismail Samani was the Samanid Amir of Transoxiana and Khorasan. His statue stands on a huge red marble stairway pedestal flanked by two cross-legged lions. Standing in front of a monumental arch topped by a large gilt crown, he holds a gold-leafed coat of arms of the country.

Erected in 1999, this 25m (82 ft) tall monument honours the 1,100th anniversary of the Samanid Empire.

---

[36] There is no difference in dating between BCE and BC or CE and AD, just in the terms. Anno Domini (AD) is Latin for "in the year of the Lord" referring to the birth of Jesus Christ. CE, the abbreviation for "Common Era" is used to mark time in the same way. BC stands for "Before Christ"; BCE stands for "Before the Common Era".

Tanzania
2008

Kilimanjaro

I turned 70 in 2008 and felt it was probably a good time to cut back a little on my travelling. Besides, I had decided to have a go at climbing Mount Kilimanjaro. But I wasn't for giving up work completely. I had Missions to Belarus, Georgia and Tajikistan, to organise judicial training programmes for UNICEF.

Few names have such a special resonance for me as Kilimanjaro. It echoes like a spell, evoking images on a par with Kathmandu, Timbuktu or Machu Picchu – remote, inaccessible, mystical. It signifies somewhere special. The origin of the name is uncertain but, after my experiences, I tend to favour a Swahili derivation which might loosely be translated as Devil Mountain – the kind of devil responsible for creating coldness.

When I came to considering what to do to celebrate my 70th birthday my thoughts turned to that magical, mystical mountain and I asked: "Why not?" My sister Una, always ready to support my dreams, asked: "Why not indeed?"

One of the world's highest volcanoes and the world's highest free-standing mountain, Kilimanjaro is a powerful visual symbol and a quintessential African image. At 5,895 metres (19,340 feet) it is Africa's highest mountain. It is described in mountaineering circles as "moderately easy" because it is possible to reach the summit without any technical climbing ability. And yet 36% of those who attempt it fail to reach the top.

We researched what equipment we would need, carefully chose our tour operator and considered which route to the summit was most appropriate for us. We opted for the Machame Route which is considered more difficult than the more popular Marangu Route (generally known as the "tourist" route). We decided to allow ourselves seven days in order to facilitate proper acclimatisation.

We were giving ourselves every opportunity for success – to the extent possible. The unknown factor, over which we had no control, was how we would react to altitude. Of course, it would have been possible to prepare by training at altitude but financial considerations prohibited this.

Una and I were well aware that getting to the top would demand mental preparation as well as physical fitness. We began our training some 18 months ahead of the event. We started trekking in the Mournes (Northern Ireland's highest mountains), in Donegal and Kerry (highest mountains in the Irish Republic), in Wales and in Scotland to get used to trekking in difficult conditions and to build up stamina and endurance. We augmented this with work in the gym, as well as swimming and cycling.

Our Training Schedule

We started off by climbing Slieve Donard, in the Mourne Mountains, on 24 April 2007. It is not so high at 2788 feet (850 metres). But it is a tough enough climb when you are not used to mountains. So, it was a good place to start. We were back again on 5 May and spent a total of 8.5 hours walking. We really enjoyed it.

Our next two days training were spent in Donegal climbing Mount Errigal (2,467 feet – 752 metres) and Muckish (2,185 feet – 666 metres). We had no eventful happenings on Errigal but we ran into some difficulties on Muckish. I had climbed Muckish a number of times in the past and I suppose familiarity breeds contempt. We decided to take a shortcut back down. Once we started, we realised it wasn't such a good idea as we were scrambling over large, loose scree but we kept going. The route we had chosen was the direct way down but the level of difficulty meant that we lost time instead of gaining it. We made it back to the car after about 5½ hours.

I was 69 on the 31 August 2007. We decided to celebrate by climbing Mount Snowdon, the highest mountain in Wales at 3,560 feet (1,085 metres). It rained all day and we were soaked after the first couple of hours. When we got to the top the mist was so thick that we couldn't see more than six feet in front of us. It was still pouring rain, was blowing a gale and we were freezing. We took one rain-soaked photograph and headed back down again. We made good time despite the rain – just six hours for the round trip. We were soaked to the skin.

Even the money in my wallet was saturated. We were glad to get back to the B&B to get changed and get warmed up.

On 28 September 2007 we climbed Carrauntoohil (3,415 feet – 1,041 metres) in Co Kerry. It was a beautiful day, better than any we had seen that summer, and we set off in bright sunshine. We headed for the 'Devil's Ladder' which is a steep rocky gully set into the mountain. The name gives an indication of the level of difficulty. But we knew it would be difficult on Kili so we wanted to be prepared. We had lunch at the top and then began our descent – even more difficult for me with arthritic knees. We made it back down the ladder without problems.

On 1 January 2008 we headed for Tyrone to climb Mullaghcairn. At 1,778 feet (542 metres) it hardly merits the title of mountain. But I believe that climbing should be fun. This was our own county; the scenery is beautiful and it was a good way to get back into training after the Christmas break.

On 20 February we decided to walk the towpath between Belfast and Lisburn, a distance of about 12 miles. We set off from the city hall at 10:20 am, stopped for a picnic halfway and arrived in Lisburn around 3 pm. We had lunch in a cafe with outside seats so that Lucy could join us. Then we caught the 4 pm bus back home.

We generally spend Easter birdwatching in the Highlands of Scotland. This year we thought that mountain walking in the Cairngorms would improve our fitness levels. However, it started snowing the night we arrived and continued to snow every night until we left. As the temperature hovered around 0°C most of the time we didn't do much mountain climbing but we managed a couple of walks on the lower slopes. The first one was a path we had followed on a previous trip and we were able to follow the route for about an hour and a half but then the snow became so deep we couldn't see the path anymore so we turned back.

The second day we walked up the ski slopes as the sun was shining and we hoped to see a Ptarmigan, a mountain bird which had eluded us so far, despite our many trips here. We were almost at the top of the mountain when the low cloud began to settle in and we decided to turn back. Just then, a group of climbers on the ridge above us disturbed a single Ptarmigan which flew over and landed close to us. We didn't even need our binoculars to see him. It was a wonderful sight; one we had been waiting to experience for at least five years.

The Belfast City Marathon is always held on the May Bank Holiday (the first Monday in May). We had planned to do the 16-mile walk but the choice that year was either the full marathon or nine miles. We entered the nine-mile walk which we completed in 2½ hours. We then decided to walk home, which meant we walked a total of about 16 miles altogether, and I still had enough energy to go out and cut the grass.

On 15 May we climbed Slieve Donard again. We managed the round trip in about five hours – a marked improvement on our previous time.

On 26 May we were in Scotland to tackle Ben Nevis. The weather was beautiful – bright sunshine and hardly a cloud in the sky – very unusual for Ben Nevis. At 4,808 feet (1,344 metres) "The Ben" is the highest mountain in Britain or Ireland. It would be our toughest test yet. The path was very steep. Just before the halfway point it changed to a very rough stone path and the going was really difficult. It would be easy to slip here. We made it to the summit in 3¾ hours and we were very pleased. The views were fantastic. Hardly a cloud in the sky and we could see for miles. We had lunch and headed back down again. The going was difficult as the stones kept slipping under foot. We made it down in four hours. We were both very tired next day but still managed a ten-mile walk through Glen Nevis. We enjoyed the walk as we listened to the birds singing. We heard a cuckoo calling and managed to locate him – a rare sighting of a bird which used to be the symbol of spring.

Time seemed to be passing very quickly now with the big day drawing ever closer. One of Una's colleagues at work was raising money to go and build houses in Africa. He organised a hike in the Mournes on June 28. We decided to join in because the money was going to a good cause. Everyone had been very generous to us in our fundraising efforts. Besides, it fitted in with our training schedule. I was able to keep up with the majority of the group, all of whom were considerably younger than me.

The final assault on Kili is generally during the hours of darkness with the aim of being on the summit to witness the sunrise. We had never climbed using head torches and needed the experience. So, we headed back to Kerry on July 19 to have a go at climbing Carrauntoohil again. Our aim was to climb it twice within 24 hours – first during the day and again at night. We set off at 10:30 am with a small group and two guides. It was cold, windy and wet so we were pleased to make it back to the B&B for 6 pm. We just had time for a quick shower before meeting the group for dinner.

We were up at 2:20 am. We set off walking with Niall (our guide) and the group at 3 am. We headed to the Gap of Dunlow and then branched off and started to climb. It was dark and the rain had started again. At the top of the hill, we had to cross a bog which took about an hour. Our head torches were working fine and we were not having any problems. Once dawn broke, we didn't need the head torches anymore. We climbed the first of the MacGillicuddy's Reeks. We were back to the B&B at 9:30 am just in time for breakfast.

That was all the major training we had planned before the big climb. We continued to do long walks at the weekends and went to the gym twice a week. We stopped all the training two weeks before we left for Kili to give our bodies time to relax before the big trek.

Kilimanjaro Here We Come

When preparations began, October 1, 2008 seemed a million light years away. Suddenly we were on the plane on our way to Tanzania. We landed in Nairobi, Kenya, and changed planes. We were met at Kilimanjaro airport and whisked off to our hotel. In the afternoon we got a briefing from our two guides and we were all set to begin our adventure next morning.

The Agony and The Ecstasy

We left the hotel at 9 am for a one-hour drive to Machame Gate. Registration proceeded at African pace – the Kilimanjaro mantra of "Pole, Pole" (slowly, slowly) applies across the board and not just to climbing. Eventually we began our trek at noon.

The temperature at the base of the mountain was about 35ºC. This had dropped to about 30ºC at the Gate, but it was still quite hot for trekking. We were climbing through dense rain forest but we paid little attention to the flora or the fauna as we concentrated on the climb. We stopped for lunch after about two hours and were suddenly reminded that we were in a rain forest. The heavens opened and it poured on us for the rest of the day. Five hours and 20 minutes after leaving the Gate we arrived at our first camp site exhausted and soaked to the skin. Our expensive coats which had been guaranteed waterproof had let us down. With nowhere to dry anything all we could do was wrap our wet clothes in plastic bags and store them away.

The porters, the unsung heroes of the mountain, had our tent set up for us when we arrived. It was a bit small and difficult to get into with aching limbs and arthritic knees but at least we were out of the rain. The porters had also set up a mess tent and had hot tea/coffee and popcorn prepared for us minutes after our arrival. Shortly afterwards we were called for supper. We were constantly amazed at the quality of the food the chef was able to turn out in such primitive conditions. He prepared a different high-quality vegetarian meal every mealtime – meals which would put some of our top restaurants at home to shame.

We had been warned about the infamous "drop toilets". Easily located by the smell they were something of an affront to one's sensitivities – but needs must. And yet these were so much more hygienic than the "flying toilets" which are commonplace in many of Africa's densely populated cities. It was worth reminding ourselves that one of the reasons we were climbing Kilimanjaro was to raise money in order to provide safe drinking water, basic sanitation and hygiene education for those in greatest need.

The night was cold and frosty. I had not seen so many stars since I was a child some sixty years ago. There was little time to admire the beauty of the night. It was necessary to get some rest. However, unaccustomed to life under canvas, we slept little. The dawn chorus was a welcome reveille. We rolled up our sleeping bags and packed our things. The porters provided a bowl of warm water for a cursory wash and a second bowl to wash one's teeth. When this was done breakfast was ready. We were being encouraged to drink up to three litres of water a day to avoid dehydration and advised to eat heartily in order to maintain our energy. But already at 3,000 metres we were beginning to feel the effects of the altitude and our appetites were not as keen as the guides, or the chef, would have liked. We ate what we could and then had our first medical. The guides measured our oxygen saturation level and our pulse rate. There was no cause for concern and we were on our way at 8 am.

This was a tough, tough day – worse than anything we had attempted before. The ascent was inexorably upward over very difficult terrain. It reminded us of our training on the "Devils Ladder" on Carrauntoohil except that this seemed to go on forever. We started the day in warm sunshine but the sky quickly became overcast and a thick mist descended, limiting visibility. Once again, we bypassed the usual viewpoints – taking photographs was pointless in these conditions.

We stopped for lunch about noon but the altitude was clearly affecting my appetite. I felt a little nauseous. I didn't eat anything. I hadn't been eating any energy bars either, which are deemed compulsory eating to keep energy levels up. In a repeat of the previous day the rain began to fall so we quickly pulled on our waterproof trousers and not-so-waterproof jackets. The heavy rain made walking difficult and we got no respite until we crawled into our tent pitched at 3,800 metres – close to Shira Caves.

Despite starting off stiff and sore we had made good time. The estimated time for the climb that day was 5 hours. We made it in 5 hours and 10 minutes. I noted that I had not been going to the toilet as frequently as I would have anticipated. Urinary retention is another problem associated with altitude sickness.

Day three began with our usual medical check. Una's pulse was a little fast and her oxygen saturation level not ideal but we were given clearance to continue. We were warned that it would be a long hard day with an estimated 7 hours hiking in front of us. This was our extra acclimatisation day and we were following the advice of the experts who say: "climb high, sleep low". Starting at 3,800 metres we were to climb to the Lava Tower at 4,600 metres before dropping down again to camp for the night at 4,025 metres. Both of us were experiencing headaches, nausea, breathing problems and urinary retention. This day was intended to help us get used to the altitude and thus ease these problems.

It was a long, long haul up to the Lava Tower. The climb was not as technically demanding as the previous day, but it was extremely tiring because we were tired starting off, because it was a long hard day (7 hours 40 minutes) and, of course, because of the altitude. It was overcast and misty but the threatened rain did not materialise. Arriving at the camp dry helped to ease the pain of aching limbs and throbbing knees.

The altitude appeared to be affecting Una more than it was affecting me. Her head was sore and she didn't want to eat. By the time we reached camp she was feeling really ill. She struggled into the tent and lay down. I brought her some soup and coaxed her to take some. The guides were not too concerned at this point.

We woke up on day four to beautiful sunshine. The camp was in a magnificent setting – the finest so far. After our usual medical the guides told me that my pulse rate was improving but didn't comment on Una's. None-the-less we set off in high spirits to tackle the Barranco Wall. It was described in the brochure as "a mere 300 metres high – not technically difficult but tiring and hard". It looked like a vertical cliff face to us. Still the indefatigable porters were getting up it with 20-kilo loads on their back. Surely, we could make it with our much lighter loads. We made it in about two hours only to be faced with a drop into the Karanga Valley, which looked like the wall in reverse. A fall here would have had disastrous results so it was best to focus on keeping in the footsteps of the guide and to ignore the potential dangers. We eventually made it to Karanga camp at 4,025 metres, exhausted and completely worn out.

The original intention was to stay here for the night and then to have a fairly easy three-hour climb on day five to the final camp before the summit. The assault on the summit is generally made during the hours of darkness in order to reach the summit in time to witness the sunrise. However, our chief guide was proposing a change of plan. I was stopping frequently to catch my breath and liked to sit down regularly for a few minutes (I would later find out that I had a heart problem and was in need of a quintuple by-pass). The guide explained that this would not be a good idea when the temperature could be - 20°C and a strong wind could make it feel much colder. He suggested that, instead of stopping at Karanga camp for the night, we press on to the final camp. Then, after a good night's rest we could make our attempt on the summit during daylight hours when the temperature would be considerably higher and the air easier to breathe.

The plan sounded good in principle so, after a hot meal, we pressed on towards Barafu camp (the final camp before the summit). We arrived there some three hours later, totally shattered. Una was feeling quite ill and more or less collapsed into the tent. It was dark and a strong wind was driving sleet and snow. Our tent was pitched in the midst of some large rocks which had become slippery with the sleet and we were quite a distance from the toilets. I prayed for urinary retention that night.

On the morning of the fifth day, we were up at 6 am as usual. We washed, had breakfast and then our medical. The news was unexpected. Una's pulse rate was 131. After a night's rest it should have been in or around 100. Anything over 120 sets the alarm bells ringing. At 131 there was only one outcome. Una had to get back down the mountain and fast.

I got the all clear. Dawson, the chief guide, said he would accompany me to the summit. He was quite confident we could make it in eight hours. Stephen, the assistant guide, would accompany Una back down the mountain. Una was keen that I would carry on and try to make the summit. She insisted that she would be OK.

I said "No". We had planned this adventure as a team effort from the beginning and had trained together from day one. We came as a team and would leave as a team. I would not make an attempt on the summit. I would accompany her back down the mountain.

I was always concerned that the trek back down would be at least as difficult as the climb – and so it proved. Still, we managed to keep up a steady pace and reached Mweka camp by early afternoon.

All the leading texts say that a quick descent will probably cure AMS (the dreaded Acute Mountain Sickness or altitude sickness). Usually there are no after-effects. Most people make a full recovery. We had dropped some 1,600 metres and Una was showing signs of improvement already. She was able to eat her lunch and had more to eat for dinner than she had had since day 1.

This was the best-equipped camp we had seen. There was running water and hand basins in which to wash. There were sit-down toilets. They were still "drop" and not "flush" but it is hard to explain the relief of being able to sit down instead of squatting when your legs are aching and will scarcely bend.

I didn't sleep so well that night – probably because my knees were so sore – and was glad to hear the dawn chorus and the porters stirring. We got up, had our breakfast and then our medical. Una got the all-clear.

Dawson advised that it would be a comparatively easy descent to the Gate. He hadn't taken into account the pain in my knees. Besides, some stretches of the path were difficult to negotiate – especially long stretches of blue clay which was slippery and dangerous.

It was a beautiful sunny day with birds singing everywhere as we negotiated the rain forest. It would have been a pleasant hike if we had not been so tired and sore and had the time to stop and enjoy the flora and the fauna. As it was we were glad to finally reach the gate after four difficult hours.

It was a little depressing to bump into erstwhile colleagues whom we had seen on the way up, now all bright and bubbly after having successfully summitted. It seemed at first as if everyone had made it except us.

Later a different picture emerged when we bumped into a South African group whom we had got friendly with on the way up. It transpired that one of the group was a lady from Bangor in Northern Ireland, who had emigrated to South Africa and made her home there. She asked me whether I knew Terry Neill who used to play football for Northern Ireland. I said "Yes – an excellent footballer". "Well,", she said, "he is my brother". It's a small world.

There were seven in the group – all super fit with a mixture of marathon runners, long distance walkers, some who participated in the triathlon and so on.

Three had not made it to the summit. One of the ladies got up the morning of the final assault to discover that her tonsils had swollen dramatically – one was the size of a golf ball. She couldn't continue. Tony was making good progress towards the summit when he just passed out cold. When he came to, he was being dragged down the mountain wearing an oxygen mask. Wimpey was also making good progress when suddenly he began frothing at the mouth and could see nothing but a white haze around him. He too was rushed down the mountain wearing an oxygen mask. The four who continued witnessed two other casualties – one being stretchered down the mountain and the other being dragged down. These reports really brought home the fact that there is a 36% failure rate, and several deaths each year, and underlined the fact that Kilimanjaro is not a mountain to be trifled with.

When we eventually got through the registration formalities our transport was waiting to bring us back to our hotel. It is hard to explain what it is like to stand in a hot shower and wash away a week's grime. It was a very forceful reminder of just how much we take for granted – clean drinking water, hot showers and flush toilets.

Our efforts had raised £6,000 for WaterAid. We know that they were deeply grateful for the money, which went towards providing safe drinking water, basic sanitation and hygiene education for those in greatest need. We, in turn, were grateful to our many sponsors who supported us so generously. We hoped that they would feel we had done them proud even though we did not reach the summit.

That was only part of our fundraising efforts. We also raised £2,000 for our little cousin Meabh who suffered brain damage at birth and was in need of a special walking frame to help her learn to walk. There were times going up the mountain when I thought I just couldn't take another step. Then I would think of Meabh and how much she would give to take just one step unaided. When the going got tough I thought of Meabh. She inspired me to keep going.

Before leaving Kilimanjaro, we wanted to have a quick visit to Moshi – a city of some 200,000 people on the lower slopes of the mountain. As soon as we arrived, we were surrounded by about 20 people, all trying to sell us something. If we went into a shop to escape, they were waiting for us when we came out. We didn't really enjoy our walk around the city. We asked our driver to take us back to the hotel.

The sun was shining brightly next morning as we drove to the airport and the rays were glancing off Kilimanjaro's snow-covered summit. At a distance it seemed ethereal, almost supernatural in its beauty. It rose majestically from the surrounding mists, towering above the plains below, oblivious to the dramas enacted on its steep slopes. For me it had lost none of its mysticism. It was, and is, a special place.

Before coming to Kilimanjaro, we had decided that, whether our efforts were successful or not, we would have earned a few days break in Zanzibar to relax. Our flight to Zanzibar took about one hour, followed by a thirty-minute drive to our hotel – the Fumba Beach Lodge. Our accommodation was a beautiful, secluded chalet right on the beach, complete with sunbeds and hammocks – the perfect place to relax. The hotel had its own swimming pool but I had always wanted to swim in the Indian Ocean. Now I had the opportunity to swim several times a day. The only difficulty was the sharp coral but it was easy to spot the coral, as the water was crystal clear.

We didn't spend all the time lazing around. We spent some time birdwatching and saw Woodland Kingfishers, Blue and Yellow Sunbirds, Hornbills, Swallows, Plover, Heron, Egrets, and Whimbrel. We also had a troupe of monkeys dancing on the roof in the early hours of the morning and waking us up. We saw giant millipedes – more than six inches long. We also saw a beautiful red starfish while swimming. And, perhaps best of all, we saw some beautiful sunsets.

We took a trip to Jozani Forest Reserve where we saw several kinds of monkey and had an opportunity to walk through the mangrove swamps – on elevated walkways. We also visited Stone Town, the flourishing centre of the spice trade and of the slave trade in the nineteenth century. We took a 30-minute boat trip out to Prison Island where the slaves were held until they could be shipped to America. The only inhabitants nowadays are the giant tortoises, originally brought in – probably from the Seychelles. The biggest tortoises are about six feet long and weigh about 550 lbs. Giant tortoises can live for hundreds of years and the eldest of those on Prison Island are estimated to be at least 200 years. It is likely therefore that some of the tortoises we saw were part of the original group brought in from the Seychelles.

We were reluctant to leave the Fumba Beach Lodge because we enjoyed it so much but we had decided in advance to spend the last few days in the Mbweni Ruins Hotel in Stone Town. Our apartment there had a balcony overlooking the pool and the beach. We visited the Old Fort and did some shopping. Una wanted some more presents and I needed a new pair of swimming shorts. We had left our gear out on the line to dry the night before leaving Fumba Lodge. In the morning my shorts had gone – someone must have been desperate and didn't fancy a one-hour drive into Stone Town to buy their own.

The birdwatching was excellent. We saw Paradise Flycatchers, Common Bulbuls, Greater Blue-eared Starlings, Java Sparrows, Sandpipers, Pygmy Kingfishers, Variable Sunbirds and Tropical Boubous. We spent some time on the beach watching crabs excavating homes for themselves. It was amazing to see them emerge from a tiny hole carrying a claw full of sand, dump it some distance away, and return to dig some more.

We also saw Indian House Crows which are now regarded as pests and are the subject of numerous control orders – including trapping, shooting and poisoning. In 1891 the Indian Government sent 50 crows to clear up waste in Stone Town. They did clear up some of the waste but they had a major impact on indigenous birds – attacking small birds and eating their eggs. Some species of small bird are now seldom seen. They also take chickens and ducklings from farmyards and are accused of attacking small farm animals. The Finnish Government funded a major control project in the early 1990s which cut the crow population by some 75%. But funding ran out in 1995 and the crow population soared again.

The time passed very quickly and it was soon time to leave. As we headed for the airport on 17 October, I remembered a conversation I had with one of the locals. He wanted to know what the weather was like in Belfast. He told me that it sometimes gets cold in Zanzibar in the winter and everyone dons scarfs and gloves. When I asked him, what was regarded as "cold" in Zanzibar he said: "25°C or under". I imagined that the temperature was likely to be a little below 25C when we got back to Belfast on 18 October. Still, the warmth of the welcome we would get from Lucy would make us feel that it was good to be home.

Thailand

The Bridge Over the River Kwai

Another place I dreamed of visiting "someday" was *The Bridge over the River Kwai*. The bridge was part of the notorious Burma-Siam railway, built by Commonwealth, Dutch and American prisoners of war. This was a Japanese project driven by the need for improved communications to support the large Japanese army in Myanmar. During its construction, approximately 13,000 prisoners of war died and were buried along the railway. An estimated 80,000 to 100,000 civilians also died in the course of the project, chiefly forced labourers brought from Malaya and the Dutch East Indies, or conscripted in Thailand and Myanmar. I was delighted when the opportunity arose, during a visit to Thailand, to go to see the famous bridge.

I was surprised to discover that the story told in the book, and later in the film, is largely fictitious. The author had never been to Thailand and had not done any meaningful research. There is no bridge over the river Kwai! The story is based on the building in 1943 of one of the railway bridges over the Mae Klong River at a place called Tha Ma Kham, five kilometres from the Thai town of Kanchanaburi. The release of the film caused the Thais some problems. Thousands of tourists would be coming to see the bridge over the River Kwai, but no such bridge existed. Tourists would feel cheated if they were taken to see *The Bridge Over the River Mae Klong*. The Thais came up with a simple solution. They changed the name of the river. It would henceforth be known as the River Kwae Yai ("Big Kwae" – to distinguish it from the original, and smaller, River Kwae – or Kwai).

Despite the fact that the bridge is not quite what it seems I found the visit very moving as I walked over and back thinking of the tens of thousands of people who died of hunger and disease in the construction of the bridge and of the railway – the deaths were real enough.

Kanchanaburi – The "Romance" Of Elephant Riding

Elephants played such a central role in films I had seen about Thailand that I thought I might get an opportunity to go elephant trekking on one of my visits to the country. I read that it was possible to go elephant trekking in Kanchanaburi. So, I stopped off there on the way to see the famous bridge. I didn't have time to go trekking but maybe a ride around some ancient ruins?

I thoroughly enjoyed the experience – until I got off again! Then I noticed that some of these great animals had holes in their forehead. I wondered what had caused the wounds and was about to ask someone when suddenly it dawned on me what the sticks were which the mahouts held in their hand. These were *"elephant goads"* or *"bullhooks"*. This is a tool employed in the handling and training of elephants. It consists of a hook attached to a two- or three-foot handle. The hook is inserted into the elephant's sensitive skin, either slightly or more deeply, to cause pain and induce the elephant to behave in a certain manner. I found it hard to comprehend how these gentle giants were treated in such a cruel way. I could not condone such cruelty. If I had spotted the wounds before I went for the ride I would not have gone. So much for the "romance" of elephant riding! I resolved never to go elephant riding again.

Just after I wrote this (13 April 2019) I read an article in the New York Times:

At Egypt's Tourism Gems, Animal Abuse Is an Ugly Flaw.
The rampant mistreatment of horses, camels and donkeys at major attractions like the pyramids of Giza has prompted calls for visitors to boycott rides.
Declan Walsh, NYT, April 13, 2019

The campaign group *People for the Ethical Treatment of Animals*, or *PETA*, called on tourists to boycott all working animals at Egypt's major tourist sites. I agree with Ashley Fruno, PETA's director of animal assistance programmes that "Such abuse has no place in modern tourism". "Boycott" was my reaction when I became aware of the abuse of elephants in Kanchanaburi. But, as the NYT article notes, things are never that simple. For example, what constitutes "ethical tourism" is a matter of debate, even among animal rights groups. Animal rides provide a livelihood for thousands of impoverished families, and some groups argue it is better to reform the owners' abusive ways rather than shun them entirely. Another major international animal rights group, the Brooke organisation, encourages tourists to be vigilant of animal abuse instead of boycotting the rides—urging visitors to watch out for signs of malnutrition and to refuse to ride with an owner who whips, or otherwise abuses, his animals. Rather than shunning the owners, Brooke tries to cajole them into better behaviour, often by appealing to their pockets.

On reflection, and after reading the NYT article, I would support the Brooke approach. Not all the elephants I checked in Kanchanaburi had holes in their forehead. So, clearly, not all mahouts used a goad to control their elephant. Why should the innocent be penalised with the guilty? Follow the Brooke advice. Avoid owners whose animals show signs of malnutrition, neglect or abuse. Reward the owners who treat their animals well by giving them your business.

Tunisia

2010

At the close of the IAYFJM's XVII World Congress, 2006, in Belfast, Tunisia had expressed an interest in hosting the Congress in 2010. I had been at several seminars in Tunis and I knew some of the people involved. I was asked if I would be the liaison person between them and the Association.

The Tunisians had chosen an excellent theme – *United in Diversity: Common Law, Civil Law and Islamic Law. Finding common ground in child protection and juvenile justice.* I applauded them for their choice of theme and said I felt it had potential for an outstanding Congress. I suggested that they might consider working on a case study to show how a child might be dealt with in the various jurisdictions. I said it was important for the world to see that we all had the same priority – the best interests of the child. The Congress was fixed for 21-24 April 2010.

Everything was going according to plan in Tunis. A volcanic eruption in Iceland was of little interest. No one anticipated that the volcanic ash ejected during the Eyjafjallajökull eruptions would result in the largest air-traffic shutdown since World War II. Flights to and from Europe were cancelled. Millions of passengers were stranded not only in Europe, but across the world. The initial shutdown over much of northern Europe lasted from 15 to 23 April, just as delegates should have been on their way to Tunis for the Congress. It is impossible to know how many delegates were stranded, or simply cancelled their flight. I didn't make it to Tunis until 23 April. I estimated the number in attendance to be in or around 200. With many speakers and workshop presenters failing to turn up everyone pitched in and covered to the extent possible. There was a continuous shuffling and reshuffling of the programme. Things went remarkably well in the circumstances. And everyone I spoke to said it had been a good Congress.

Top of my list of places to visit in Tunisia was the archaeological site of Carthage. Carthage played a central role in antiquity as a great commercial empire. During the Punic wars, the city occupied territories that belonged to Rome. This led to its destruction by the Romans in 146 BC. The Romans then rebuilt the town on the site of the ancient city. Apart from its exceptional historical importance (it is a UNESCO World Heritage Site), I was interested in Carthage as the home of Hannibal. I took the opportunity, on two separate occasions, to visit the site which is just a short bus ride from Tunis. With so much history to absorb, it would merit further visits.

When in Tunis, I liked to spend some time wandering around the souks looking for souvenirs. Most of the shops and boutiques date back to the 13th century. They are arranged in a maze of narrow streets and alleyways – some only a metre wide, guaranteeing it is a traffic-free zone. It would be easy to get lost in the medina, but I never felt nervous or apprehensive. The stall holders were all so friendly and polite. Of course, they encourage you to stop, view and buy their wares – but not in an aggressive way. I always enjoyed my time there and found it a good way to relax.

Turkmenistan
2011, 2012

In August 2011, I had my first mission to Turkmenistan. Turkmenistan is a country in Central Asia, bordered by Afghanistan to the southeast, Iran to the south and southwest, Uzbekistan to the east and northeast, Kazakhstan to the northwest and the Caspian Sea to the west. It possesses the world's fourth largest reserves of natural gas and has substantial oil resources. Although it is wealthy in natural resources, in certain areas, most of the country is covered by the Karakum (Black Sand) Desert. It was part of the Russian Empire from late in the 19th Century until 1991 when it was granted independence. It is about 89% Muslim, 9% Eastern Orthodox (+ 2% undeclared).

After independence, Turkmenistan's Chief of State, Saparmurat Niyazov, replaced communism with a unique brand of independent nationalism reinforced by a pervasive cult of personality. After he was elected President in 1992, he spent much of the country's revenue on extensively renovating cities, Ashgabat in particular. Ashgabat is now quite a nice city with wide, treelined boulevards. It has a population of about one million. Most of the buildings are sparkling white and everywhere is spotlessly clean. There are endless magnificent statutes *of* and monuments *to* the President. In 1999 he declared himself president for life. He conducted frequent purges of public officials and abolished organisations which he deemed threatening. He died of cancer in 2006. The Deputy Prime Minister, Garbanguly Berdimuhamedov, was named interim head of government and won the special presidential election held in early February 2007. He was re-elected in 2012 with 97% of the vote. I understand that most of the opposition was in prison.

I arrived in Turkmenistan at 3 am on Sunday 14 August 2011, after a six-hour flight from London. The temperature was 29°C even at that hour of the morning. Turkmenistan is four hours ahead of Belfast. It was 4:15 am (12:15 am our time) before I got to bed. I had the Sunday off to recover. It was very hot – I guess it was over 40°C. I didn't feel like going out. I had booked into the Grand Turkmen Hotel and had a nice balcony which was shaded from the sun and overlooked the swimming pool. But it was too hot even to sit out there. I took a quick run out to visit the local market, which was five to ten minutes' walk away. I wanted to buy a large bottle of water and a few apples. That cost me a mere $1. If I had bought the same water in the hotel, it would have cost me about $5 – and no apples included.

I took a few photographs when I was out – and ran into trouble with the police again! I took some photographs of the hotel, including the swimming pool. Then I took a couple of photographs in a beautiful little park just across the road. There is a really nice building across the main road from the hotel, so I photographed it. Immediately a policeman ran up to me and said: "No photograph!" I said: "It is a beautiful building". He replied: "Bank. No photograph." and crossed his arms to form an "x". He wanted to see the photograph and said: "Delete". So, I did and then showed him the other photographs. He said: "OK". As I walked off, he called me again and said: "Photograph us?" I walked back to where he was standing with a colleague who had arrived. I took their photograph and showed it to them. They smiled and walked off. So, we parted as friends! Or did we? Perhaps I was fooling myself. I wonder if they were savvy enough to know that you can recover "deleted" photographs on the computer unless you overwrite them with fresh pictures. In any event, it was hard to know what could be photographed and what was forbidden.

I had three Missions to Turkmenistan. The first was in August 2011 in the capital, Ashgabat. The second, in July 2012, was in Turkmenabat, about one hour's flight to the east. The third, in August 2012, in Dashoguz, about one hour's flight to the north. Turkmenistan is divided into five provinces and the intention was that I would cover the remaining two provinces in July and August 2013.

Turkmenistan is very hot in the summer. The highest recorded temperature in Ashgabat is 48°C. The highest in Turkmenistan is 50.8°C. On my first two visits the daytime temperature was in the low 40's. On my third visit, the temperature hit 47°C. I didn't go out a lot but I wanted to at least see something of the capital. So, on my third visit when my driver collected me from the airport and dropped me off at my hotel, I asked him to call back for me about 3 pm and take me to see some of the tourist sights.

When I met him at reception, he told me he was not taking me out. The temperature was in the mid 40°s C. I wouldn't enjoy it. He called back in the evening and we had a quick tour of the Independence and Peace Monument, the Arch of Neutrality, the Ertugrul Gazi Mosque, the Ashgabat National Museum of History and the Carpet Museum. But perhaps the most imposing was The Mausoleum & the Spiritual Mosque of Turkmenbashy, the largest mosque in Central Asia. Twenty thousand men and women can pray at the same time. The mosque was named after the first president of Turkmenistan, Saparmurat Niayazov (known also as Turkmenbashy – the Leader of Turkmen). The floor of the mosque is covered by handmade Turkmen prayer mats and an enormous eight-sided carpet decorates the very centre of the mosque. There is an amazing echo when you stand in the very centre and clap your hands. The mosque consists of two floors – men pray on the first floor, women on the second. It has four minarets and a golden dome located in the central upper part of the mosque. The height of each minaret is 91m. The dome is 50m high and covered in gold. The mosque is surrounded by beautiful fountains and gardens.

There is a magnificent mausoleum of the first President and his family located next to the mosque. The presidential tomb is in the centre surrounded by those of his father who died in World War II and his mother and two brothers who died in the 1948 earthquake. The cult of Turkmenbashy continues in death!

I enjoyed my time in Turkmenistan. The judges were accustomed to being told what to do and could not contemplate using their initiative when the legislation left room for manoeuvre. Apart from that I found the judges in Asghabat and in Dashoguz easy to work with. They seemed to enjoy role play – once they got the hang of it. I found it almost impossible to engage with the judges in Turkmenabat. They listened politely and one of them proposed a vote of thanks to me at the end. I discussed the matter with the UNICEF rep who told me that the people in that particular province didn't like "outsiders" coming to tell them what to do. "Outsiders" included fellow Turkmen from Ashgabat!

UNICEF invited me back in 2013 for missions in the final two provinces. Those two Missions did not take place. The Turkmen Government had been happy to have UNICEF fund a whole range of projects – for example children's homes, detention centres and the like – but were unwilling to allow visits to any of them, to see how the money was being spent. I understand that UNICEF grew tired of their stonewalling and pulled the plug.

For my part, I found the Turkmen people I worked with very friendly. I enjoyed my time with them. They all insisted that I come back and see them. I got an invitation to go trekking in the Karakum desert in December 2012. The Turkmen Karakum is approximately 135,000 square miles in area. It extends about 500 miles from west to east and 300 miles from north to south. The idea was to take part in a camel safari. I was asked whether I had ever ridden a camel. I said I had ridden a camel across the Sahara. They said that if I had ridden across the Sahara, I would have no trouble with the Karakum – how many weeks had it taken me? I said it had taken me approximately one hour because I had only crossed a tiny little corner of the Sahara close to Cairo. They then realised that I was joking about my expertise at camel riding. But they still wanted me to give it a try. I said I didn't think I would last even one day, let alone two weeks. It would have been a wonderful experience but the invitation came just a *few* years too late.

UK

1982-1993

In my introduction (page 15 of the current text) I say that I have visited 75 countries, at least once, in my role as a lay magistrate. If I try to calculate the number of actual visits to any individual country, no country comes close to the UK. I was the Northern Ireland representative to the BJFCS for ten years, the last two of those as Chair. This meant four or five trips per year to London giving a total number of visits of 45-50. The International Association met in London four or five times over the years bringing the grand total to between 50 and 55. Switzerland would come a distant second, the USA third, Brazil fourth, Argentina fifth, South Africa sixth, Canada seventh, France eighth, Italy ninth and Myanmar tenth.

Readers might find it strange, therefore, that the UK report is one of the shortest in the book. It would be surprising if I had not done some sightseeing in and around London over the years and if our colleagues on the International Association had no desire to take a tour of one of the world's great cities. However, I point out in the introduction, that I am not writing a travel guide. If there is nothing I can add to what is already available in travel books I don't include it here.

The Isle of Man

In July 1991, I organised and led a delegation from the BJFCS to the Isle of Man to study their system. Birching was their standard punishment for delinquent youth. But birching had just been outlawed by the European Court of Human Rights and I wanted to see how they were getting on without it. I found a certain nostalgia for the Birch even though it had been demonstrated that it wasn't really a deterrent. Young offenders from Liverpool, Manchester and other cities were reported as regarding it as a badge of honour. They felt good to be able to say they had been birched and that there was nothing to it.

The Isle of Man claims to have the oldest continuous parliamentary body in the world – Tynwald Court. Each year, Tynwald Court participates in the Tynwald Day Ceremony in the Royal Chapel – St John's. After a religious service in the Royal Chapel, the members of Tynwald process to Tynwald Hill, one of the ancient open-air sites where the people could observe their parliament in action. By statute, each Act of Tynwald must be promulgated on Tynwald Hill within eighteen months of enactment or it ceases to have effect.

Any person may approach Tynwald Hill on Tynwald Day and present a Petition for Redress. If the Petition is in accordance with the Standing Orders, any Member of Tynwald may subsequently request that the substance of the petition be considered. Matters are redressed by this simple but ancient procedure which can lead directly to the enactment of legislation.

Following the proceedings on Tynwald Hill, presided over by the Lieutenant Governor, the members of Tynwald return to the Royal Chapel where a formal sitting of Tynwald takes place. Any Acts promulgated on the Hill that day are certified at the sitting in St John's. Our visit coincided with Tynwald Day and we were honoured when invited to participate.

I want to mention briefly one other event I organised as Chair of the BJFCS. In May 1993, I jointly organised a conference with Dean Lou McHardy, Dean of the National College of the NCJFCJ, in Reno, Nevada. The conference was an overview of the American system specifically for the BJFCS. I led a delegation of 25 members to the conference. You will find a report on the trip below (p251).

USA

1981, 1982, 1987, 1988, 1993, 2003

Liam and I arrived in New York on June 25, 1981. We stayed at the Barbizon Plaza Hotel overlooking Central Park. Since we are both keen birdwatchers our first thought was to head for the park. We saw a queue lined up close to one of the park gates and assumed we needed tickets to enter Central Park so we waited in line. It turned out the queue was for subway tokens – we kept ours for souvenirs! When we figured out where the actual entrance to the park was, we walked around for a while. Suddenly the heavens opened for a torrential rainstorm. We had never seen rain so heavy, or felt rain so warm. We were soaked to the skin in minutes – before we could even begin to look for shelter. We tried sheltering under the trees, but that made little difference. We had to wait until the water stopped running out of our clothes before we could go back into the hotel. Since then, Central Park has become our metaphor for any particularly spectacular downpour!

We did the usual tourist things. We visited The Statue of Liberty. We took a very interesting day-long tour which included Chinatown, Harlem, the World Trade Center and Liberty Island. The next day we took an all-too-short but spectacular helicopter tour of New York. We planned to visit Macy's but it was closed when we got there. We walked around Times Square and along 5th Avenue to visit St Patrick's Cathedral. But our favourite spot was Central Park, despite the soaking we got on our first visit. In my many visits to New York since then, I always made time for a walk in Central Park.

We flew from New York to Toronto on June 28. I have covered this part of our trip in my report from Canada. (p65). We flew from Toronto to San Francisco on June 30. Toronto had been hot (mid 30sC) and I was looking forward to sunny California.

I completely forgot that San Francisco has its own microclimate and is famous for its fog, especially in the summer months – June, July and August. I understand it took early explorers a little while to realise that there was a bay there! Travel guides advise that "if you plan on visiting the Bay Area any time between June and August, make sure to pack a sweatshirt, jeans, and warm layers in case you get caught in the fog." We arrived on one of those damp, chilly days at the end of June. The temperature was just 11°C!

We had booked into the Bedford Hotel, but our room wasn't ready. We went out looking for a coffee shop. I was dressed in light summer trousers and a mesh-type, short-sleeved shirt. I remember clasping my coffee cup with both hands to try and thaw them out.

The "must do" list for San Francisco included riding on the Cable Cars, visiting Fisherman's Wharf and walking down Lombard Street. Muir Woods is just a short coastal drive of less than an hour from downtown San Francisco, over the Golden Gate Bridge and through the scenic Marin Headland Hills. We couldn't miss the opportunity to see some of the oldest, tallest, most amazing trees in America. Liam had read about trees 300 feet tall and couldn't wait to see them.

But perhaps most important of all, we wanted photographs of the Golden Gate Bridge. The American Society of Civil Engineers declared the bridge to be one of the Wonders of the Modern World. At the time of its opening in 1937, it was both the longest and the tallest suspension bridge in the world. The entry in Fromer's travel guide described it as: "possibly the most beautiful, certainly the most photographed, bridge in the world."

The fog didn't lift during our stay and we had no chance of getting a decent photograph of the bridge. This was not what we expected in "Sunny California". Luckily, we did find blue sky and warm sunshine when we crossed the Bridge on the way to Muir Woods.

Not far from the northern end of the Golden Gate Bridge we passed through the small town of Sausalito. It struck me as little more than a village at that time – a picturesque residential community. It was close to San Francisco (just four miles from the city centre) but far enough away to be free from the city's famous fog. I remarked to Liam that, if I had a choice of places to live, Sausalito would be No 1.

Monterey is about 120 miles south of San Francisco – a two-hour drive. We both love wildlife and Monterey Bay is known as the Serengeti of the sea. I suggested we go there. The Bay is home to many species of marine mammals, including sea otters, harbour seals and bottlenose dolphins; as well as being on the migratory path of Gray and Humpback Whales and a breeding site for elephant seals. Killer whales are also found along the coast, especially when grey whales migrate, as they hunt the whales during their migration north. Many species of birds, and sea turtles, also live here. Several varieties of kelp grow in the bay, some becoming as tall as trees, forming what is known as a kelp forest. So much to see and so little time. There was one species I particularly wanted to see – the sea otters.

Sea otters are a common sight along the Monterey coast as they inhabit the nearshore kelp forests. They rest within the kelp, anchoring themselves by rolling in it until it is wrapped around them so that they won't float away from family and friends while asleep. Their diet is mainly sea-stars, crabs, urchins, or abalone which live in the kelp. When they catch their prey, they bring it to the surface in open water, away from the kelp. Then, floating on their back, they use their stomach as a table. They place their prey on their "table" before proceeding to rip it apart with their forepaws or crack it open with their powerful teeth.

Even more amazing is the way they use rocks as anvils to crack open mussels, clams and crabs. Sometimes they will use a stationary rock at the water's edge and hammer the shell against it. Sometimes they will return to the surface with a clump of mussels, roll over onto their back, lay the mussels out on their "table" and then crack them one by one on a stone they have placed on their chest. Such fascinating animals – we could have watched them all day. But we had run out of time.

Monterey was the only place we visited in the US that Liam didn't like. His reason for not liking Monterey – when we decided to have lunch, we couldn't find a restaurant which served French Fries!

We left San Francisco on Saturday July 4 and flew to Los Angeles where we stayed at the Ambassador Hotel[37], Wilshire Boulevard. I found Los Angeles large and sprawling with no real centre. Our hotel seemed miles from anywhere. We went out walking, hoping to locate a Catholic church with a view to attending mass. We hadn't realised that no one walks in Los Angeles. There was no one about to ask directions and we seemed to have been walking for miles without locating the church. Suddenly I spotted two people walking on the opposite side

---

[37] where Robert Kennedy was shot in 1968

of the road. The traffic was light enough for us to make our way across and we ran to catch up with the other two pedestrians. Consider my surprise when they turned around and I recognised my milkman from back home. He and his wife were also out looking for the church. We had a good laugh and continued our search together. Eventually we found it. The church was called St. Basil's. After mass we had a drink with our milkman and his wife at our hotel. We didn't get to see any Fourth of July celebrations except on TV – although we did hear the fireworks.

Top of the list of places to see in LA were Disneyland and Magic Mountain but, first, we had some books to deliver to a Sheriff's Department Officer, Deputy Sheriff Jack Howard. He found us before we had an opportunity to go looking for him. He came to our hotel to get us on our first full day in LA. We stopped briefly at his home to pick up his wife and son before going to Disneyland. The next day they took us to Six Flags Magic Mountain. We struck up an immediate relationship with Jack, his wife, Jeanie, and their son Denny. Jack told Liam what it was like to be a police officer in LA. He explained that he must carry a gun at all times even when off duty, or risk being fined a week's wages.

We kept in touch with the Howards for a number of years. Jack was a big, strong man, while Jeanie was very frail. It came as a shock, therefore when we got a letter from Jeanie to say that Jack had died suddenly. We kept in touch with Jeanie for another couple of years, until she stopped responding to our letters. We assumed that she had gone to live with her son in Oregon. We didn't have his address.

We had set one day aside to take a bus trip to San Diego so that Liam would have an opportunity to visit Mexico. San Diego shares a 24 km border with its sister city Tijuana, which has been nicknamed "The Gateway to Mexico". It is the most visited border city in the world. Current estimates are that more than fifty million people cross the border between these two cities every year. (For more detail see p142, under Mexico, Tijuana).

On the way back we stopped briefly at the famous San Juan Capistrano monastery, perhaps best known for the annual "Return of the Swallows" which is traditionally observed every March 19, (The Feast of Saint Joseph).

In 1982, Liam and I rounded off a 14-day holiday in Canada with three days in Boston. We booked into the Copley Square Hotel on August 3. We concentrated on some interesting tours where Liam learned a lot about the

Boston Tea Party and the American Revolution. When we arrived at the airport to catch our flight home, we found that a computer malfunction was causing chaos with too many people booked on our flight. We were offered a free weekend in New York if we gave up our seats. Liam was disappointed that we had to refuse but he knew that his mum was impatiently awaiting our return.

In 1987 I got an invitation to attend the 50th Anniversary Conference of the NCJFCJ in Cincinnati, Ohio. My wife died in 1986. I didn't want to leave Liam so soon after losing his mum. I asked him if he would like to come with me. We would have five days in Cincinnati. I also planned to visit courts and juvenile facilities in seven other US cities. We would visit New York, Cincinnati, Washington, DC, Baltimore, Houston, Los Angeles, Las Vegas and Minneapolis (in that order) and finish off with three nights in Montreal. Liam was excited just thinking about it. We began our tour in New York on July 9 that year.

The New York Family Court was headed by an Administrative Judge (Judge Kathryn A McDonald, at the time of my visit) one Supervising Judge for each county and 42 Family Court Judges. The Administrative Judge was responsible for administering not only the Court itself but also the building (cleaning, heating, repairs etc), and employing staff.

The Supervising Judge also had to review cases heard by other judges. This was done randomly as an aid to ensuring uniformity. But it was also done by way of an appeal where someone was unhappy with a decision. During my visit Judge McDonald was reviewing a case which was causing some controversy. Indeed, a headline in the New York Post that morning (July 10th 1987) read:

12-YR-OLD KILLER FREED. He stabbed pal, 14, with a steak knife.

The boy had been released on probation the previous day. The report was quite scathing in its attack on the judge who should, according to the New York Post, have handed out the maximum sentence in the circumstances – 18 months in jail.

The reporter had, in my view, paid scant regard to the details of the case, highlighting instead the few points which supported his case. Judge McDonald's task was to review all of the evidence before the court and decide whether or not the decision reached by the judge was the correct one in the circumstances. She kindly provided me with a transcript of the case to read in my own time – I could not be present while she reviewed the case lest it was alleged that I was party to the decision. I left New York without hearing what decision Judge McDonald came to. On my reading of the case the disposal was appropriate in the circumstances.

While Judge McDonald was reviewing the case outlined above, I sat in one of the courts observing the proceedings. A number of cases were "B-Petition: Fact-Finding Hearings" (B-Petitions refer to the termination of parental rights). In most cases the children were already in care and many of the parents could not be traced. Evidence of search was presented. Where no contact had been made within the last six months the child was assumed to have been abandoned. Where parents could be traced evidence of service of notice of the hearing had to be presented. Once this evidence was presented to the court a date was fixed for the "Dispositional Hearing".

A number of "O-Petitions" were also heard (wives seeking exclusion orders for alleged ill-treatment). In one case the judge refused to issue an exclusion order against a husband of eleven years but did order him to be of good behaviour.

In another where the youngest son was 32 the judge also refused an exclusion order but offered to set a date for a hearing within five days.

A "V-Petition" (custody and visitation) raised some interesting questions. The mother of a two-year-old child alleged abuse by the father against the child. She had moved to Pennsylvania raising questions of (1) jurisdiction – should the case be heard in Pennsylvania where the mother was now residing or in New York where the child was born and where the father was residing and (2) access – since the father had a long way to travel.

The father wanted to have the child with him in New York all weekend. The mother objected strongly on the grounds of the alleged abuse (which he strenuously denied). When the father's wish was not granted, he then asked for permission to visit in Pennsylvania. It was eventually agreed that the child would be brought to his mother's house in Pennsylvania and that he could visit there all day on Sunday on condition that either his mother or father were present at all times.

I was particularly impressed in this case with the amount of time and effort the judge devoted to attempting to reach a compromise. He would call the parties together, make suggestions to them, send them off separately to talk it over with counsel, and then bring them together again. When agreement was eventually reached, he asked them to try to communicate more. It was his belief that poor communication was at the root of the problem.

Baltimore

Baltimore differs from New York in that it has a Juvenile Court and not a Family Court. The Court has one judge, (Judge Mitchell, at that time) and eight "Masters" (deputy judges appointed by the judge). A "Master" sitting in court has the full powers of a judge with the exception that anyone who is unhappy with the decision reached may make an immediate appeal to the Judge.

The Judge will listen to all the evidence, recorded on tape, read the transcripts and decide whether or not the case should be heard again. Appeals are not common and only a small percentage of those appeals which are made are upheld by the Judge.

The workload of the Masters is very heavy e.g., Master Joyce Mitchell (no relation to the Judge) whom I sat in with hears 30+ cases per day. For this reason, efforts have been made to streamline the system by having the Masters specialise in the cases they hear. Master Mitchell dealt almost exclusively with neglect and abuse. While I was there, there was a heated discussion with social workers who had gone, in error, to one of the other Masters seeking a Place of Safety Order. Instead of referring the case to Master Mitchell, as would have been normal procedure, he heard it and turned the request down. The social workers were very angry and wanted Master Mitchell to hear the request again. She said she could not do that. The other Master should not have heard the case, but they should have known the proper procedure. It was not within her power to ignore the other Master's ruling and hear the case again. She suggested they might approach the Judge.

The dirty linen all came out as the social workers argued that the Master who heard the case was incompetent – he never wrote his reports up as he should, he was always late – one report being months overdue and so on. Master Mitchell protested that she shouldn't be listening to this and advised them again to make their complaints to the Judge. They stormed off without, as far as I am aware, making their appeal.

In New York (and many other States) "minors" become "adults" once they reach the age of 16. 16-18-year-olds caught pushing drugs in New York felt the full weight of the law. Many young people got round this by taking a bus to Baltimore or Philadelphia to push their drugs. Here they were minors until they were 18 so that if they were caught the consequences would be much less severe.

I spoke to a judge from Philadelphia who was quite angry that this should be so and wished the law changed so that he could ship the pushers back to New

York where they belonged. "Let them deal with them, it is their problem not mine."

Judge Mitchell took a very different view. "People will always work the system," he said. "We must accept that. Who knows, a harsh approach might make them into hardened criminals. Maybe the way we deal with them will get them back on the straight and narrow. Wouldn't that be in everyone's best interests?"

During our visit to Houston, I had the opportunity to join Judge Andell on the Bench. He invited the prosecutor and defence attorneys to approach the Bench and explained who I was. He asked whether they had any objections to me joining him on the Bench while he heard the various cases. There were no objections. I found it a great experience.

One case of interest concerned a 15-year-old found by the police to be in possession of a gun. When asked by the judge where he had got it, he said he had stolen it. He went on to declare himself in need of the gun for self-defence. The youth came from a deprived background – a very poor home in a rundown area. Despite this he had an excellent school record. His grades indicated that he was university material.

The judge told me he had never seen such good grades for a young person from that background. He gave the boy a very stern talking to, laying the options before him – university or detention centre. He asked for an undertaking that the boy would concentrate on his studies. Giving the boy an absolute discharge, he said: "You were one stupid act away from being dead. Get out of here. I don't want to see you in here again". I was very impressed with the judge's handling of the case.

One of the senior judges in Houston, Judge Lowry, invited Judge Andell, Liam and I to spend the weekend at his holiday home in Galveston (on the Gulf of Mexico). We were delighted to accept the offer. Fishing was a true passion for Judge Lowry. He and Judge Andell planned to spend most of the Friday night fishing for sharks. I am not a fisherman but they insisted I try my hand at it. The craic[38] was great all night and the time passed quickly. I was delighted that I had no "luck" in catching sharks but the others were much more successful. Liam managed to catch an eel which he threw in again. The Lowry and Andell families

---

[38] "craic" is an Irish word meaning fun, entertainment, and enjoyable conversation.

joined us for breakfast. Liam and I were fully integrated and really made to feel included. I must admit that I enjoyed barbecued shark for breakfast.

Back in Houston Judge Andell took us to watch his son playing in a Little League (Baseball) Competition. Later we joined the judge and his friends and tried our hand at baseball. I regarded myself as competent at rounders, but baseball was, literally, a whole new ball game. My innings was quickly ended.

The Judge kindly arranged a VIP tour of the Johnson Space Center in Houston for us. Our guide was a senior Air Force officer. We got to see some off-limits areas where visitors are not normally allowed. Among other things, we got to sit at the Flight Director's desk (most visitors are on the other side of a screen) and to see a 50-foot deep swimming pool where astronauts train. We were told that, being in the water, with the upthrust making them float, is somewhat akin to the sensations felt in space.

Judges in Texas are elected and Judge Andell's term was coming to an end. We happily donned T-Shirts bearing the slogan "Vote Judge Andell" and joined him in a run around the neighbourhood. Judge Andell teased us that Texas would be the high point of our trip. It turned out to be true! But LA had some surprises in store for us too!

My strangest experience in Los Angeles was my visit to Juvenile Hall. Juvenile Hall was a high security prison holding young people who had been found guilty of the most serious offences including murder, attempted murder and rape. It was built to hold some 1200 young people but was housing almost double that number at the time of my visit. All the guards were armed and the tension was so high that I felt the place could explode at any moment. The guards walked around as if they were tracking dangerous animals and, I guessed, that was how they regarded the young people.

One of the Directors approached me and said he had received an unusual request. A group of young people had asked if they could speak to me. The Director warned that these were the most dangerous of a dangerous lot – all murderers or rapists or both. He would fully understand if I refused. I said I would be happy to talk to the boys. I was escorted into a large cage-like enclosure. Two guards armed with rifles stood either side of the door. I sat down at a table with another two guards armed with rifles – one standing each side. Strangely I didn't feel afraid at all. I was simply horrified at the way these young people were being treated. Was it any wonder if they acted like animals? That was how they were being treated. It was hardly surprising that the tension was so high. The only way it was being contained was by arming the guards to the teeth. Why bother thinking about rehabilitation? These young people had been sentenced to life without parole. I talked to the group for about 30 minutes. I found them easy to talk to. They simply wanted to know how young people charged with offences similar to theirs would be dealt with in Northern Ireland. What would the penalty be? Did we have life without parole? I left feeling a great sense of sadness at the way these young people were being treated. There was no doubt that they were guilty of the most heinous crimes. But no account was taken of the fact that they were children. They were given no hope of redemption. There was nothing to look forward to except death.

Our main judicial contact in LA was Judge Gabriel A Gutierrez. When we visited his court a young offender who had been failing to attend school and presumably getting up to further mischief was appearing before him. The judge gave him a stern warning that, if he didn't mend his ways, he faced the prospect of ending up in Juvenile Hall. He concluded: "Believe me, you don't want to go there". I could have added: "Amen to that".

Judge Gutierrez arranged for us to visit a children's home. I was disappointed to find out that social services in California, far from leading the world, were as hamstrung by inadequate funding and lack of resources as our own back home.

Before ending the visits to LA's Juvenile Courts, I was in for another surprise, something I would never have imagined possible. I got an invitation to a wedding.

One of the Judges, Judge Jamie Corral, had been dealing over a number of years with a young girl who was deaf and unable to speak who had suffered appalling abuse at the hands of her parents and had been placed in care. Judge Corral had taken a fatherly interest in her case at the many judicial reviews of her placement in the children's home.

At some outing she had met a boy her own age who was also deaf and unable to speak. They had fallen in love but the rules of the Children's Home did not allow the boy to visit her in the Home. She appealed to Judge Corral who arranged regular meetings. They wished to get married and she asked Judge Corral if he would marry them in his court and allow them to hold their reception there as he was the only "family" she had.

This situation had never arisen before but permission was granted. Liam and I were privileged to be present at this unique occasion. It was good to see two young people so happy and to feel part of a system which had helped to bring that happiness about.

LA had turned out to be another high point of our trip and we still had to visit Disneyland and Universal Studios where a lot of Liam's favourite series were made – The Incredible Hulk, Battlestar Galactica, Buck Rogers, The A-Team, Knight Rider. Much to Liam's delight, we were briefly waylaid by some Cylons[39] on arrival at Universal Studios. He was thrilled to see both the A-Team van and the Knight Rider's car! We also saw a live "Miami Vice" show with real explosions – we could feel the heat!

Finally, Judge Gutierrez and our other contacts arranged for Liam to speak with a senior LAPD officer about possible future job opportunities in the Los Angeles Police Department. We left LA on a high note.

After Los Angeles we headed for Las Vegas to stay with another of my judge friends – Judge McGroarty – and visit his court. We also visited some of the casinos where we found that you could have free breakfast if the management thought you were going to stay and gamble. Of course, we tried our hand at gambling. I lost several dollars. Liam managed to lose only $1.05.

We hired a car and drove to the Grand Canyon. On the way back we visited Phoenix, Arizona. I was amazed to spot a Road Runner – a bird which can fly but spends a lot of time running along the road after insects – just as its name implies. Up until that moment I had thought of Road Runners as cartoon characters.

Minneapolis was the next stop on our schedule. My good friend, Judge Lindsey Arthur, who had arranged contact with the various judges we had visited, lived there. Unfortunately, we would not be able to meet with him as he had retired just before our arrival and he and his wife were off visiting their daughter. However, he had spoken to one of his colleagues, Judge Albrecht, who would be happy for us to visit his court.

As juvenile court hearings are normally held in camera, we were privileged every time we visited a court. A defence attorney in Minneapolis initially objected to our presence, but Judge Albrecht persuaded him it was OK to have us there. The attorney was a public defender, i.e., appointed by the court for people who can't afford an attorney, and we had an interesting discussion with him afterwards about the various aspects of that job.

---

[39] robots from Battlestar Galactica.

When we met with Judge Albrecht he enquired where we were staying. We said we had booked into a Holiday Inn on the outskirts of the city. He insisted we come and stay with him. It would be more convenient and he could show us around the city. In particular, he would like to take us to see the Minnehaha Falls Regional Park one evening.

This park takes its name from the beautiful stream and its spectacular falls that plunge 53 feet into a gorge before running out to the Mississippi River. The poet Henry Wadsworth Longfellow gave the waterfall national fame in the *Song of Hiawatha*, which I remembered reading at school. The epic poem is much too long to remember word for word but I did remember snippets. I remembered how "…the Black-Robe chief, the Pale-face, With the cross upon his bosom…" had been welcomed by Hiawatha and how the strangers replied: "Peace be with you, Hiawatha, Peace be with you and your people". It saddened me to think that this was the beginning of the end for the native Indians' way of life.

But these melancholy thoughts quickly evaporated as we walked towards the spectacular falls. Trails lead in all directions. It is possible to descend into the gorge below the falls via a staircase to get a great view of the falls from below or walk along the river to the confluence of Minnehaha Creek and the Mississippi River. Unfortunately, we didn't have the time. I thanked Judge Albrecht for taking us there. I said that if we lived in Minneapolis, we would be regular visitors to the park, especially since it had all necessary facilities including a restaurant and an off-leash dog park. I told him it was the highlight of my visit. It was a highlight for Liam also, but not *the* highlight.

I had told Judge Arthur that Liam was considering submitting an application to join the LAPD. He had a word with the local Chief of Police and arranged for Liam to do a night patrol in a police car with one of the officers (they don't always work in pairs) from the Minneapolis Police Department[40]. He couldn't have come up with a better present for Liam.

Florida

Before leaving Cincinnati, I had been invited to attend the NCJFCJ's next Annual Conference which was to be held in Fort Lauderdale in July 1988. Liam

---

[40] Police Departments, and Police Officers, are responsible for policing within city limits, as in the NYPD or the LAPD and, here, the MPD. Policing outside the city line is the responsibility of the Sheriff's Department and Deputy Sheriffs.

came with me. Once again, I had an opportunity to visit Courts and children's facilities but we also fitted in some sightseeing. I hired a car for two weeks because the places we planned to visit – Orlando, Titusville, Fort Lauderdale and Miami – were a considerable distance apart.

We spent three nights in Orlando where we visited Disney World. My favourites were the Tiki Room and a boat trip to an island full of tropical birds.

We visited the futuristic EPCOT Centre (Experimental Prototype Community of Tomorrow), where there were a number of pavilions representing different countries and staffed by their respective nationals. We were surprised to see that they mistakenly categorised the Republic of Ireland as being part of the UK. There were mock-ups of Italian and Chinese monuments. We had our first ever Japanese meal at EPCOT that evening.

We spent a couple of days in Titusville, near Cape Canaveral, staying with new-found friends, Lynn and Carl. We had met them at the 1987 Cincinnati conference, during an evening cruise on the Ohio river. Liam had told them about his all-night patrol in a police car. So, Lynn and Carl arranged for him to ride around with a deputy sheriff in Cocoa Beach, near where they lived.

Carl put me on the spot when he invited me to view his gun collection. He was so proud of it. It was an impressive collection – 32 rifles ranging from guns for hunting to heavy-duty military guns. I do not support the way in which the Second Amendment[41] of the US Constitution is interpreted, bearing in mind the number of people killed in mass shootings in the US each year. But I did not want to raise that debate with my host. I restricted myself to asking questions for clarification: Do gun collectors require a licence? Is there any restriction on the number or type of guns collected? Are any checks made to ensure that large collections of weapons are properly secured?

Carl worked as a structural engineer at Cape Canaveral and gave us a guided tour. Unfortunately, there were no space launches scheduled during our visit.

After the conference in Fort Lauderdale, I was given the opportunity to visit the "Last Chance Ranch" a last-ditch effort aimed at rehabilitating young offenders before they went to jail. Liam said it reminded him of a comment made by Sherlock Holmes in one of the books he had read[42]: "Make him a jailbird now, you make him a jailbird for life". Our guide was a young man who had been

---

[41] A well-regulated militia, being necessary to the security of a free State, the right of the people to keep and bear Arms, shall not be infringed.

[42] "The Blue Carbuncle"

found guilty of second-degree murder, though he claimed to be a victim of a misunderstanding. He was exquisitely polite as he showed us around! The young residents had many disincentives to try to run away. The ranch was miles from anywhere and was surrounded by swamps and forest. There was a high probability of close encounters with alligators or rattlesnakes!

We finished off with a few days in Miami. On one of those days, we drove down the Florida Keys to Key West, the southernmost point in the continental USA. It was a memorable drive.

We also managed to squeeze in a day trip by boat to Freeport on Grand Bahama Island. Our stay on the island was so short that it was hardly worth the effort of getting there. We had a grand total of two hours, just enough time to visit a souvenir shop and go to the beach for a quick swim! It turned out that we would have been better paying a little more to fly to Nassau from Miami, as we could have had a full day there. Still, it was a nice way to round off our trip to Florida. One curious thing we noticed: people drive on the left (the Bahamas being a former British colony), but all the cars are American with left-hand drive.

After Florida I was appointed an Honorary Member of the NCJFCJ – the first non-American member of the American Association! I got a standing invitation to attend the NCJFCJ Annual Conference each July, sometimes as a speaker, to chair a workshop or simply as a guest.

The most abiding memory of my visits to courts in numerous States in the USA is of the judges rather than courts and cases. I was not sure what to expect of judges who are elected. I suppose I pictured them as some kind of politicians. As it turned out I found them to be very politically active but political with a small "p". They carried heavier workloads than we would be expected to deal with in Northern Ireland. In most instances they were at work by 8:30 am (8 am in Judge Andell's case), sometimes working until 6:00 pm.

As Juvenile Court judges most of them had to carry the burden of administration – dealing with hiring and firing of staff, heating, cleaning, repairs and so on. One judge told me that if he gave up juvenile work, he would get rid of all the administration, cut his case load in half, retain the same salary and have a better chance of promotion to greater things. Why did he continue? It was essential, he said, that those judges dealing with young people should genuinely have their interests at heart. And that was certainly true of the judges I met.

They were all involved in all kinds of schemes to provide alternatives to custody, to raise money to sponsor such schemes – several of the younger men regularly competed in marathons to raise money, they were haranguing politicians, talking to the media and so on.

Their Association is also deeply involved in researching effective methods of dealing with dependency cases, children in conflict with the law and incorrigible youth. They are in the forefront of drives for reforms in the legal system. This is what I mean by political with a small "p". The juvenile court judges I met were all at the hub of the juvenile justice system.

I found the American Courts, and the cases they dealt with, not all that different from our own. Disposals, too, were pretty similar but there were some interesting non-custodial disposals. I would have liked, however, to have been able to bring home some of the enthusiasm, some of the dedication, some of the initiative of the judges I met. I was deeply grateful to them for welcoming me in their courts.

Reno, Nevada

In May 1993, I jointly organised a conference with Dean Lou McHardy, Dean of the National College of the NCJFCJ, in Reno, for members of the BJFCS. I was Chair of the BJFCS at that time. I first met Lou at the World Congress in Rio in 1986 and met him each year since then at the NCJFCJ Annual Conference, as well as at the World Congress in Turin. We had become close friends. I led a delegation of 25 members to the conference. My sister Una said she would like to come with me.

The Reno trip started off badly when the travel agent with whom I had booked flights for myself and Una went bankrupt. I thought I was OK as I could claim compensation from ABTA only to find that the travel agent concerned had not paid his insurance and his membership of ABTA had lapsed. Nowadays ABTA would step in anyway, but not then. We just had to rebook our flights. We then had a second mishap but this time, thankfully, a minor one. We arrived in Reno to find that our baggage had failed to come with us. However, it was located in San Francisco and was delivered to our hotel the next day.

One of the group members was Dilly Gask, immediate past Chair of the BJFCS. She was accompanied by her husband, John. Una and I knew them well and had visited them at their home in Devon. Neither John nor Una was attending the conference, so they were great company for one another. They had plenty of time to explore the city and the surrounding countryside.

Una and I had arranged to spend a few days after the conference with another friend from the NCJFCJ – Jim Toner. Jim's grandparents were Irish so we had plenty to talk about. Jim's youngest daughter, Bridget, had a beautiful horse which she called Sean and, of course, we had to have a go at riding him. He was a big horse, about 16 hands high, and seemed quite placid as I rode him round the paddock. Then it was Una's turn. Una had never been on a horse before and was a little nervous so Bridget was holding the halter. I asked her to let go for a moment so that I could get a photograph. The paddock gate was open and something spooked Sean. He took off through the gate at full gallop and disappeared into the sunset with Una clinging on for dear life. I guessed Una wouldn't be able to hold on for too long and my heart was in my mouth as we went looking for her. Sure enough, she had fallen off when Sean went down a steep incline on the other side of the road. Luckily, she had no major injuries, just scrapes, bruises and broken glasses. We went to the hospital for a check-up but she got the all-clear. Sean had not gone too far and returned home of his own accord.

From Reno we went to Disneyland and then to Las Vegas to spend a couple of days with another judicial friend and take in one of the marvellous shows. We rounded off our trip with a visit to San Francisco. I wanted Una to have a ride on the Cable Cars, walk down Lombard Street and dine on Pier 39, Fisherman's Wharf. While walking around the Fisherman's Wharf area we noted people queuing on Pier 33 to catch a ferry to Alcatraz. We are both keen birdwatchers and had seen the 1962 film: *The Birdman of Alcatraz*. We knew that Stroud had become an expert on bird diseases in Leavenworth Prison, not Alcatraz, and that he had written his book there. He wasn't allowed to keep pets when transferred to Alcatraz. But the film stimulated an interest in the prison and we had subsequently watched *Escape from Alcatraz*, the 1979 film based on the escape of Frank Morris and the Anglin brothers in 1962. It was only 15 minutes away by ferry. It would be a shame to miss this opportunity.

First, let me tell you a little about the island.

Alcatraz Island is located in San Francisco Bay, 1¼ miles offshore. When first "discovered", the island was home to countless Californian brown pelicans. The Spanish called it "La Isla de los Alcatraces" (Alcatraz is an archaic Spanish word meaning "pelican"). The Spanish name, shortened to "Alcatraz", was popular and was accepted as the official name.

The island was developed with facilities for a lighthouse, a military fortification and a military prison. A high security Federal prison was added in 1934. The infamous and notorious federal prison, was home to the likes of Al Capone, George "Machine Gun" Kelley, and Robert "The Birdman" Stroud – offenders regarded as too dangerous to be housed anywhere else. Despite being so close to the city of San Francisco it was isolated from the outside world by the cold, strong, tremendously fast-flowing currents in San Francisco Bay.

There were 14 separate escape attempts, involving a total of 36 prisoners, between 1934 and 1963, when the prison closed. Nearly all were caught or didn't survive the attempt. On June 12, 1962, the guards found that three prisoners were not in their cells: John Anglin, his brother Clarence, and Frank Morris had got out of the prison and off the island. Did they make good their escape or were they swept out to sea? Their fate remains a mystery to this day. The file remains open!

The prison closed in 1963 because it became too expensive to run. Today, the island's facilities are managed by the National Park Service as part of the Golden Gate National Recreation Area. It is open to tours. It attracts one million visitors per year. Una and I decided that we would make it one million and two in 1993.

Upon arrival, a National Park Ranger provided a brief welcome and orientation. We were then free to explore the island and the prison at our own pace. A 35-minute audio headset tour of the cell block was a real highlight. We listened to stories told by former inmates and prison guards as we walked around.

After the Cell Block tour, there was an opportunity to see all the sites around the island and visit the Alcatraz Museum and bookshop. We had a different agenda, we went birdwatching! I mentioned above that the island was once home to countless Californian brown pelicans. After the prison's closure in 1963 Alcatraz once more became a sanctuary for seabirds. To the best of my knowledge pelicans no longer nest here. They have been replaced by other seabirds including cormorants and pigeon guillemots, and waterbirds such as snowy egrets and black-crowned night herons.

I suspect that the majority of visitors to the island have no idea that Alcatraz is a premiere spot for viewing nesting seabirds. Most seabirds nest on inaccessible offshore rocks. The determined bird lover would need a boat and good binoculars to get a glimpse. But Alcatraz visitors can see the mating, nesting and parenting behaviours of these birds up close. When Una and I arrived there seemed to be new-born chicks everywhere with proud parents trying to keep them under control. We spent a pleasant hour observing the frenetic activity, and we didn't even need our binoculars.

Before leaving San Francisco, I wanted to let Una see one of my favourite spots in the world – Sausalito just across the Golden Gate Bridge. Of course, we also had to see the famous redwood trees in Muir Woods National Park. We headed off through the park for a walk. The heavens opened and we were soaked to the skin. Luckily, we had our cases in the car so we were able to get dry clothes out and change in the toilets. One of the rangers was upset for us and informed us that: "it never rains in Muir Woods in July". Unfortunately, there is always the exception which proves the rule!

San Antonio

I had a seminar in San Antonio, Texas, in February 2003 immediately after completing a mission to Myanmar. I tried to arrange a direct flight from Bangkok, or at least from London, to New York. However, I had different sponsors for the Yangon trip and the American trip and they would not agree. So, I arrived in Belfast late in the evening only to be up bright and early next morning to catch a flight back to London and then on to New York and San Antonio.

San Antonio is a beautiful city in south-central Texas with a rich colonial history. It is the seventh largest city in the United States and the second largest in Texas. It attracts thousands of tourists annually. But I didn't come here as a tourist and I didn't have much time to look around. Top of my list of "must see" was "The Alamo", an 18th-century Spanish mission, now preserved as a museum commemorating the infamous 1836 battle for Texan independence from Mexico. The Battle of the Alamo was a pivotal event in the Texas Revolution. A small group of Texas soldiers, supported by a small number of volunteers under the command of the legendary frontiersman, James Bowie, were besieged by around 1500 Mexican soldiers. The defenders, numbering around 200 in total, held out for 13 days. The Mexicans won that battle but the brutality shown to those who survived the siege prompted Texas settlers to join the army in droves. They got their revenge when they defeated the Mexican Army in the Battle of San Jacinto on April 21, 1836, ending the rebellion. Texas then became an independent republic before joining the Union 20 years later. I found it a moving experience standing in The Alamo trying to imagine what it was like for the defenders during those 13 days.

The only other one of the city's highlights I had time to check out was the River Walk. The San Antonio River Walk is a city park and network of walkways along the banks of the river. I checked out the shops and cafes which line the walkway in the evening after the conference ended. I was up bright and early next morning to have time for a long walk along the river and do some birdwatching. Then it was time to hurry back to the hotel for breakfast and get a taxi to the airport.

Travel Broadens Our Mind and Widens Our Horizons.
Does It Have to Be Global Travel?

Travel broadens the mind and widens our horizons by expanding our world and our relationships, smooths the rough edges of our personality and makes us a more rounded person. In getting to know more people we become more aware of self. We become more humble, more appreciative of what we have.

I was one of the lucky ones. Few people get an opportunity to travel as extensively as I have and fewer still may have an opportunity in the future. As the Covid-19 pandemic sweeps around the world like a tsunami, it is beginning to look as if globetrotting may become a thing of the past. Since its emergence in Asia at the end of 2019, the virus has spread to every continent except Antarctica. It was more than just a health crisis. By stretching every country it touches to the limit, it had the potential to create devastating social, economic and political crises that would leave deep scars.

Many of our communities were unrecognisable. Some of the world's greatest cities were deserted as people stayed indoors, either by choice or by government order. Shops, theatres, restaurants and bars were closed. Factory gates were locked. Holiday resorts around the world had empty hotels and deserted beaches. The world's greatest airports became parking lots for planes which were flying nowhere. People were losing jobs and income, with no way of knowing when normality will return. It is becoming increasingly likely that, when it does, we may not recognise it. Life will be "normal – but not as we knew it".

Scientists say that viruses of this type are circulating all the time in the animal kingdom.

They think it is highly likely that the Covid-19 virus came from bats but first passed through an intermediary animal. The hypothesis that this virus was a zoonotic[43] disease like MERS (Middle East respiratory syndrome) and SARS (severe acute respiratory syndrome) was supported by the observation that many of the early cases were associated with a "wet" food market in Wuhan, China, where numerous live animals of different species were kept prior to sale. The SARS outbreak, 2002, moved from horseshoe bats to civets before infecting humans. In the case of MERS (2012) the intermediate hosts were camels. Pangolins[44] are being considered as a possible bridge for Covid-19.

It is impossible to say what will be normal but it is possible that the majority of people will decide not to go abroad for holidays.

Some will be afraid to go because of the experience of Covid-19 and the possibility that pandemics will become more frequent as we intrude further into the natural world.

Some will not be able to afford to travel because prices will be exorbitant following the enforcement of social distancing on buses, trains and aircraft.

Should you feel resentful that Covid-19 has robbed you of the opportunity to travel, modern technology could ease your frustration. You can ease your wanderlust while still at home. Do an Internet search for information on your favourite country – view the landscape, study the culture, download recipes. Close your eyes and imagine you are in the Rain Forest in Borneo, for example, as you listen to Gibbons howling and birds calling with a chorus of crickets and cicadas in the background. Then open your eyes to watch drone footage of Vietnam's Son Doong, the world's largest natural cave.

Some people may no longer want to travel. They may have become much more connected with nature during the lockdown and have come to enjoy peace and tranquillity. They may find staying close to home much more relaxing than the hustle and bustle of travel.

If you still feel the urge to travel, there is no need to travel to the ends of the earth to broaden your mind and widen your horizons. Why not focus on your country of birth or adoption? Seize the opportunity to gain a greater awareness of your built, natural or cultural heritage. Let your moto post Covid-19 be: *Travel less but travel better!* Now is the time to discover destinations closer to home. Travel nearby, travel safe. Local can be both interesting and inspirational. Let us consider Ireland, for example.

---

[43] a disease that normally exists in animals but that can infect humans.

[44] a scaly mammal that looks like an anteater.

People come from all over the world to visit the Giant's Causeway, the Glens of Antrim, Fermanagh's Lakes, Newgrange, the Lakes of Killarney, Glendalough. And those are just a few of Ireland's world-famous beauty spots. How many of these have you visited? Are any of them within easy access?

One day, about 60 years ago now, an American tourist stopped and asked me where he was. When I said "Loughmacrory" he replied: "It must be your Tourist Board's best kept secret. I have travelled round the world and I have never been in a more beautiful place. And yet, I can't find it on my map, and there is no signpost to tell me where I am". Now that I have been round the world myself, I, too, believe that Loughmacrory, and all the more famous beauty spots in Ireland, can match anything the world has to offer. What hidden gems, like Loughmacrory, are waiting to be discovered in your own County? Perhaps even in your own parish?

And it is not just the scenery. Stand on the Hill of Tara and listen to the faint echo of "The harp that once…". Let the music in the wind sing sweetly to your soul. Walk amongst the peaceful stone ruins of Clonmacnoise and think of the scholars who came here from all over the world during Ireland's renowned Golden Age of learning. Join those who come to study the unusual flora of the Burren. Or take a ferry out to what *National Geographic* has described as "One of the world's top island destinations" – the Aran Islands, at the mouth of Galway Bay. The largest island, Innis Mór, has a World Heritage site, Dun Aonghasa, set on a dramatic 300-foot cliff edge. There are Celtic Churches of historical significance. There are lots of birds for those interested in the fauna. On my last visit, I saw four cuckoos sitting in one small clump of trees – something you would never see on the mainland. And then there was a stoat, frolicking about on a stone wall.

Some of you may be thinking – OK, so you are seeing new sights, but what about experiencing new cultures, meeting new people, people who have a different way of life, who live by a different set of rules? You will find such people within your own community. Let me give you two examples.

I was admonishing a father whose 14-year-old son was not attending school. He looked at me and said: "Master, if my son is to be a successful farmer, he needs to be able to tell the weight of a bullock, standing on four legs, to the nearest pound. You won't be teaching him that at school."

I was hill-walking in Donegal on a glorious summer's day. I saw an old man sitting on a wall, in front of a whitewashed cottage, smoking his pipe. As I approached him, I was thinking how lucky he was to live here, surrounded by purple heather, overlooking the golden strand at Downings and the sparkling waters of Sheephaven Bay. Speaking to him in Irish, his native tongue, I said: "You have a magnificent view". He replied: "Yes, it is a pity you couldn't eat it". To understand this rather strange remark it is necessary to put ourselves in the old man's shoes, to see the world through his eyes. His perspective was that of a resident farmer, trying to eke out a living and feed his family on a farm consisting almost entirely of rock, with scarcely enough soil to plant a few potatoes or to provide fodder for a solitary cow. My perspective was that of a tourist, pausing for a few moments to take a photograph to show to friends back home. As I went on my way, I reflected that I had assumed that the old man would agree with me. I had not considered that his life experiences were different from mine.

What about those who, for whatever reason, cannot physically travel at all? Are they condemned, in St Augustine's words: "…to read only a page?" What about becoming a time traveller? In the current context, with the focus on Ireland, a good friend and colleague, Dr Des O'Reilly, encourages us to explore the enormously rich placename heritage of the Province of Ulster.

Let me draw on a few examples I mentioned earlier.

In my report of the time my family lived with my Aunt Roseann in Marketcross I mention a well called "Dochaile". Generally, wells in Ireland are named and revered because people believe the waters can cure certain maladies and afflictions. But there are no records of "Dochaile" being regarded as a holy well. Local historians were unable to tell me where the name came from and were unsure about the correct spelling. I turned to local folklore.

Local people believe that Colmcille[45] was on his way to Carrickmore where he was to oversee the construction of a new monastic settlement. He made his way through Glen Upper, intending to skirt around Lough Fingrean to join the

---

[45] St Colmcille is one of Ireland's three patron saints. The other two are St Patrick and St Brigid. Colmcille was no angel. His strong personality and forceful preaching ruffled feathers and in 563 AD he was accused of starting a war between two Irish tribes. Following calls for his excommunication, the High King banished him from Ireland never to return. Together with 12 companions he established a new monastic settlement on the island of Iona.

road running from Creggan to Carrickmore. When the entourage reached Marketcross they were parched by the hot summer sun. Then, as they tramped through the heather they came, quite unexpectedly, on a spring of pure, clear water and stopped to slake their thirst. Colmcille was told that the spring was on the boundary between farms and ownership was disputed. He was asked to adjudicate. He blessed the well and said that it would never run dry. There would always be a bountiful supply of water – "deoch uile" – a drop for everyone.

When our family moved to Loughmacrory in 1948, I wondered who Macrory was and why the district had taken his/her name. The most plausible explanation came, again, from local folklore and another story about Colmcille.

Colmcille was not the most mild-mannered of saints. He was renowned as much for his quick temper as for his holiness. The paths he travelled were strewn with blessings and with curses in equal measure.

After spending some time in Carrickmore, Colmcille decided to head back to Derry. He and his entourage stopped at a farm about three miles from Carrickmore where he had been told they could get a plentiful supply of fresh spring water to see them on their way. He asked the farmer to direct them to the well. The farmer, perhaps fearing that he would be expected to feed the entourage, said that the well had dried up. Colmcille replied that if the well had dried up it would now start flowing and would never dry up again. But, if the farmer was telling a lie, the well would swell up until the entire farm was covered in water and the flood would never subside. The farmer begged for forgiveness. Colmcille's idea of mercy seemed even more harsh to the farmer. He said the flood would subside after three members of the Macrory clan drown in the lake. Some of the senior members of his entourage pleaded with him to be more lenient. He said the three didn't all have to be named Macrory – just one Macrory and two others with a "Mc/Mac" in their name. He was not in the mood for further discussion and called on his entourage to move on.

Sometime after the well was cursed, a Macrory died in the lake. Now I knew why no one would swim there – two others with "Mc/Mac" in their name would die before the curse would be lifted. My brothers and I were McCarney, our cousin was McElduff, so we were particularly at risk and were advised that we should not tempt fate. We survived. The curse remains.

Readers will recall that I spoke about a "fairy fort" called Dunnaminfin which stood on a hill a short distance from our house in Glen Upper. The name would appear to come from the Irish "Dún na mná fionn" – the fort of the fair-haired women. I could find no clues as to who the "fair-haired women" were. At the bottom of the hill, just below the fairy fort, was a small area, little bigger than a double grave and completely enclosed by shrubs. Our landlord Henry believed that it was a fairy graveyard. It may well have been a grave, perhaps dating back to famine days. It was impossible to tell without excavating the site.

Let me give you one more example. There is a stretch of road halfway between Glen Upper and Loughmacrory called "Ballybrack Road". I asked someone who lives along the road: "Where is Ballybrack? Where does the name come from?" The answer I got was: "Ballybrack is the name of the road from Milltown round to our country."

"Placenames, whether they relate to natural features, flora and fauna or land divisions and settlement patterns are not arbitrary sounds without meaning. They should be regarded as records of the past that reveal something about the places in question and, perhaps, about the people who lived there."[46]

A local historian told me that the old people didn't pronounce the name as Ballybrack. Their pronunciation was more like Bully brack. A search on "placenamesni.org" found that "Bulley brac" is an old form of the name. This suggested the name came from the Irish word 'bualie'.

Traditionally, cattle were brought to summer pastures on the mountains on May 1. The local girls would live on the mountainside churning butter while the men remained on the farm. The mountain grass was believed to be greener and sweeter, giving the butter a better taste. This place where the churning took place was called the 'buaile'.

Having answered the question "Where does the name come from?" you might wish to research the folklore around the 'buaile'. Cows, milk and butter were believed to be affected by fairy influences. Witches were particularly active on May eve and May Day – more than at any other period of the year. I will leave you to carry out the research.

Steps to outwit the fairies and witches were no longer practiced when I was growing up but the memories lived on. Luck played an important part in butter-making. Periodically, the fairies were blamed when anything went wrong such as butter that wouldn't 'break;' which was actually caused by milk not separating

---

[46] An Illustrated Guide to The Placenames Of Ulster by Des O'Reilly, 2015

properly. A neighbour who called during butter-making was always expected to say, "God bless the work" on arrival and then 'take a turn at the churn' before leaving, so as not 'to take the luck'. I never had the pleasure of making butter but was always happy to 'take a turn at the churn'.

These are just a few examples to demonstrate the possibility of a fascinating journey into the past. While searching for answers we are likely to generate further questions.

What evidence is there that Colmcille came to Carrickmore? What was he doing there? Why was he banished from Ireland? Where did he go?

What is the role of fairies in Irish mythology? Who were the Tuatha Dé Danann? What role did they play in the various colonising invasions of Ireland?

What are the possibilities that the grave at Dunnaminfin was a famine grave? It was not unusual for people to be buried at some convenient spot when there was no one left to carry the remains to the church graveyard which might be several miles away.

There is no telling where our voyage of discovery will take us as we delve into local folklore, Gaelic legend and mythology in an effort to find answers to our questions. What is certain is that our journey will traverse older cultures and provide a mode of communion with our ancient and not-so-ancient ancestors.

A World Apart
A Peek Behind the Curtain of Ignorance

This final chapter is not for everyone. In it I discuss how the beautiful, exciting, fun-filled world which we enjoy is a dark, dangerous, stressful place for others. We could travel the whole world over and not be aware that this shadowy world exists. An evil web of dirty little secrets where children are enmeshed at the mercy of ruthless abusers.

Trafficked within and across borders, press-ganged into prostitution, pornography and other intolerable forms of child labour, these children are overwhelmingly drawn from the ranks of the most vulnerable – refugees, orphans, abandoned children, child labourers working as domestic servants, children in armed conflict – and those whose sexual abuse began at home or in other familiar surroundings.

As the enormity of the situation sinks in you may feel a rising sense of anger about the abuse of children in the brothels of Bangkok, the train stations of Moscow, the truck routes of Tanzania or the sidewalks of Manila. The abuse of children may be closer to home than you think, or wish to believe. It is understandable that some may not want to peek behind the curtain of ignorance and bear witness to the dreadful acts that are hidden in plain sight. If that includes you, you can skip the rest of this chapter.

I explained in the introduction that I was not travelling as a tourist and that my role was to train judges in the use of international instruments concerning the rights of the child. My travels enabled me to gain insights into the countries visited and the inhabitants which would not be available to the average tourist. The insights we have discussed so far have, generally, been good. A few have been bad. In this final chapter I will give you a glimpse of the ugly – a shameful world where children are sexually abused. It is hard to imagine a more difficult and shocking obstacle to the realisation of human rights than the commercial sexual exploitation of children.

In 1996, I attended the 1st World Congress Against the Commercial Sexual Exploitation of Children, held in Stockholm, Sweden. The exploitation was exposed and denounced as unacceptable. The Congress adopted a clear and unequivocal position: the shameful abuse of children, so long a dirty secret, must end.

It raised public awareness of the appalling scale of the problem. Governments and community groups together affirmed that children are not property to be bought and sold. Children's rights must be respected and their voices must be heard. It was time for nations to adopt a policy of zero tolerance to child pornography and prostitution. There were calls to summon the resources and the political will to end the abuse which continued to strip countless children of their rights, their dignity, their childhood – and often their very lives.

Drawing on the strength of the UN Convention on the Rights of the Child, the Stockholm Declaration and Plan of Action was intended to be an inspiration for national plans of action and other practical measures to improve the lives of children everywhere.

Five years on (17-20 December 2001), we met again in Yokohama, Japan, for the 2nd World Congress, to see what had been done, what remained to be done, and to pledge renewed action to put an end to the unacceptable.

Attendance at the four-day Congress, sponsored by ECPAT[47], UNICEF, the Japanese Government, and the NGO Group for the Convention on the Rights of the Child was by invitation only. The 3,334 delegates were drawn from 138 countries. Apart from Government delegates there were representatives from 21 international institutions, including UNICEF, and 148 NGOs (Non-Government Organisations) from all around the world, including the IAYFJM (International Association of Youth and Family Judges and Magistrates), which I was representing.

We had children there who were brave enough to tell us what it is really like. I will give you two examples: a 14-year-old, kidnapped and forced into prostitution, raped and sodomised, sometimes at five-minute intervals, every day for six years until rescued by one of the NGOs; an 18-year-old, rescued after eight years, whose friends were all dead, having died at the hands of pimps or clients, garbage thrown on the rubbish heap, not meriting a police investigation.

We learned in Yokohama that trafficking in children and women in the Asia-Pacific region alone had victimised about 30 million people over the three

---

[47] End Child Prostitution in Asian Tourism.

decades 1971 to 2001. Carol Bellamy, (then) Executive Director, UNICEF, told us that the commercial sexual exploitation of children was nothing less than a form of terrorism – one whose wanton destruction of young lives and futures must not be tolerated for another year, another day, another hour.

Delegates to the Yokohama Congress agreed that the commercial sexual exploitation of children was a complex problem and that there was a need for more research. They called on Governments to draft policies and programmes and prepare five-year plans. They then packed their bags and went home to spend Christmas with their families, opening presents around the Christmas tree.

For the subjects of the Congress, Christmas Day was like any other – sitting huddled in a little room waiting for the next man to come in and defile them. How many more men would rape and abuse them before the delegates met again to review their five-year plans?

I warned you above that the abuse of children may be closer to home than you think, or wish to believe. The day I returned home from Yokohama (December 20, 2001) I was greeted by a headline in the local morning newspaper, which said: "Irish child slave ring exposed". The article claimed that trafficking, organised by criminal gangs, was widespread throughout Ireland and Britain.

The sexual abuse of minors is an historical phenomenon which can be found in every culture and in every society. It affects every corner of the world, from the richest countries to the most impoverished.

Both Congresses focused on the commercial sexual exploitation of children, but what about those whose sexual abuse begins at home or in other familiar surroundings?

More than 8 out of 10 children who are sexually abused know their abuser. They are family members or family friends, neighbours or babysitters, probably someone the child likes and trusts.

The abuse often occurs over a period of time which can run into years. It starts gradually through a process called grooming and continues until the full extent of the sexual abuse desired by the abuser is reached.

The primary reason that the public is not sufficiently aware of child sexual abuse as a problem is because there is a culture of silence around the issue. 73% of child victims do not tell anyone about the abuse for at least a year; 45% do not tell anyone for at least 5 years and one out of every three children will never disclose.

Incest and intra-familial abuse accounts for about one third of all child sexual abuse cases. Behind the reluctance to tell could be feelings of betrayal, of guilt, shame and confusion. The abuser may advise the child that, if they tell, no one will believe them; or that the offender will be sent to prison and they (the child) will carry the blame; or they may threaten to harm the child or family members. If the child discloses to a parent, or sibling, they, in turn, put pressure on the child not to tell anyone else because, if the story got out, the entire family would be disgraced. Should the child tell someone in authority – teacher, social worker – the family is likely to close ranks around the offender and call the child a liar.

Many abusers hold responsible positions in society. Some will seek out employment which brings them into contact with children. Some will hold positions of trust which can help to convince other adults that they are beyond reproach. If the abuser holds a position of authority – a teacher, a priest, a coach – the child may feel unable to refuse to do as they are told. The abuser can demand that the child tells no one about "their little secret" warning that, if they do, no one will believe them.

The Congress in Stockholm raised public awareness of the extent of the problem and Governments were pressured into drawing up action plans to deal with it. Yokohama went considerably further.

At the global level we had the adoption of three major treaties that address sexual exploitation and abuse:

1. ILO Convention No 182, which calls the involvement of children in prostitution and pornography one of the worst forms of child labour;
2. The Protocol on the prevention of trafficking of children and others, part of the UN Convention against Transnational Organised Crime; and
3. The Optional Protocol to the Convention on the Rights of the Child, in this case a measure aimed at ending the sale of children, as well as child prostitution and child pornography.

Many countries drew up national plans of action to combat sexual exploitation and assist victims. Measures included the establishment of special bodies to protect child rights, reform of juvenile-justice systems; training of police and judicial authorities; and all-out crackdowns on those who sexually exploit children.

There was increased police action growing out of cooperation among national law enforcement groups and Interpol. The private sector was more involved, particularly the tourism and Internet-service industries. More resources were committed on a regional basis to combat sexual exploitation. There were calls to strengthen international cooperation and action at every level of every society. Governments and media outlets were told that they must have the courage to end once and for all the shameful silence that kept commercial exploitation and abuse a secret and move forcefully to identify and bring to justice culpable individuals and criminal networks. Where police services had previously left allegations of child abuse for the parents to sort out, they were now being told to have such allegations properly investigated and offenders brought before the courts where there was evidence of abuse.

It was quickly realised that the home is not the only theatre of violence. Others, such as care homes, schools, church premises, youth clubs and sports clubs, are also environments in which episodes of sexual abuse can occur.

The raising of public awareness meant that a trickle of brave individuals came forward to report how they had been abused. When the trickle became a flood, the authorities were forced to set up inquiries.

Over the past three decades, the Roman Catholic Church has been rocked by a series of sex abuse scandals worldwide. Repeatedly, allegations about priests were dismissed by their superiors, priests were moved elsewhere and were free to abuse again.

The Independent Inquiry into Child Sexual Abuse (IICSA) in England and Wales was set up by the then Home Secretary, Theresa May, in 2015, to investigate claims against local authorities, religious organisations, the armed forces and public and private institutions – as well as sexual assaults carried out in schools, children's homes and at NHS sites. It would also investigate claims of failures by police and prosecutors to properly investigate allegations.

An IICSA report, published on July 31, 2019, found that both Nottingham City Council and Nottinghamshire County Council had repeatedly exposed vulnerable children to sexual and physical abuse, including repeated rapes, sexual assaults and voyeurism, at many council homes as well as in foster care, during the 1970s, 1980s and 1990s.

In a damning assessment of the councils' failures, the report said: "For more than five decades, the councils failed in their statutory duty to protect children in their care. These were children who were being looked after away from their family homes because of adverse childhood experiences and their own pre-existing vulnerabilities. They needed to be nurtured, cared for and protected by adults they could trust. Instead, the councils exposed them to the risk, and reality, of sexual abuse perpetrated primarily by predatory residential staff and foster carers."

Allegations of the abuse of young players at football clubs in the United Kingdom began in mid-November 2016. Former professional footballers waived their rights to anonymity and talked publicly about being abused by former coaches and scouts in the 1970s, 1980s and 1990s. This led to a surge of further allegations, as well as allegations that some clubs had covered up the abuse.

Within a month of the initial reporting, the Football Association, the Scottish Football Association, several football clubs and over 20 UK police forces had established various inquiries and investigations. By July 2018, 300 suspects were reported to have been identified by 849 alleged victims, with 2,807 incidents involving 340 different clubs.

An independent report by the Social Care Institute for Excellence, published April 4, 2019[48], concluded that abuse in the Church of England is so widespread that all parishes should print the safeguarding hotline number on service sheets.

The report states that the Church of England should have been a place which protected all children and supported victims and survivors and the inquiry's summary recognises that it failed to do this. Professor Alexis Jay, chair of the inquiry, said the Diocese of Chichester prioritised its own reputation above their welfare. The church's response "was marked by secrecy and a disregard for the seriousness of abuse allegations".

On October 6, 2020 IICSA released a report on the sexual abuse of children in the Church of England. The report stated that the culture of the Church facilitated it becoming a place where abusers could hide. Deference to the authority of the Church and to individual priests, taboos surrounding discussion of sexuality and an environment where alleged perpetrators were treated more supportively than victims, presented barriers to disclosure that many victims could not overcome.

Another aspect of the church's culture was clericalism, which meant that the moral authority of clergy was widely perceived as beyond reproach. The report states that allegations of abuse against priests were ignored, minimised or dismissed by church leaders.

The report notes that the Church forgave paedophiles after they expressed remorse and allowed them to carry on working instead of protecting children. 'Forgiveness' was used to justify a failure to respond appropriately to allegations as was 'the seal of the confessional', which creates a "duty of absolute

---

[48] The Social Care Institute for Excellence, an independent charity, carried out the review and survey on behalf of the Church of England.

confidentiality" on the information disclosed. Evidence given to the inquiry suggested that some victims may have been pressured by church workers to forgive their abuser, causing further harm and potentially bringing them back into contact with that individual.

On November 10, 2020, IICSA published an in-depth, 162-page, report into child sexual abuse in the Catholic Church in England and Wales. The Inquiry found that, between 1970 and 2015, the Church in England and Wales received more than 900 complaints involving more than 3,000 instances of child sexual abuse, made against more than 900 individuals, including priests, monks and volunteers. Those complaints involved more than 1,750 victims and complainants, though the report said the true scale of abuse was much higher and would likely never be known.

When complaints were made, the church invariably failed to support victims and survivors but took action to protect alleged perpetrators by moving them to a different parish. "Child sexual abuse," the report says, "was swept under the carpet."

Professor Alexis Jay, the chair of the inquiry, said: "For decades, the Catholic Church's failure to tackle child sexual abuse consigned many more children to the same fate. It is clear that the church's reputation was valued above the welfare of victims, with allegations ignored and perpetrators protected. ...The church's neglect of the physical, emotional and spiritual wellbeing of children and young people in favour of protecting its reputation was in conflict with its mission of love and care for the innocent and vulnerable."

The Report noted that two previous inquiries into abuse in the Church, by Lord Nolan in 2001 and Lady Cumberlege in 2007, had brought change and improvements, but their recommendations had been implemented too slowly and not in full.

The inquiry found that child sexual abuse was "far from a solely historical issue", adding that more than 100 allegations of abuse had been reported each year since 2016.

Professor Jay expressed disappointment that the Vatican's ambassador to the UK, the papal nuncio, refused to participate in the IICSA inquiry and that the Holy See would not provide a witness statement. Professor Jay commented: "the responses of the Holy See appear at odds with the Pope's promise to take action on this hugely important problem."

The Report concluded: "The Catholic Church's explicit moral purpose has been betrayed by those who sexually abused children, and by those who turned a blind eye and failed to take action against perpetrators… The Church needs to be more compassionate and more understanding of the lifelong damage that child sexual abuse can cause".

In all the examples listed above, the willingness of people "in the know" to "turn a blind eye" allowed the virus to take hold. The failure of those in authority to have due regard for the seriousness of the allegations once they were aware of the abuse, and the lack of concern for the welfare of the victims, allowed the infection to become an epidemic. Then, when their back was to the wall, the authorities prioritised their own reputation above all else.

As inquiry followed inquiry, the most recent always "the largest to date", the general public began to lose interest. Some accused the police of being over-zealous in the quest for evidence. Some believed that the pendulum had swung too far and that the alleged perpetrators were presumed guilty unless proven innocent. The focus in all these enquiries has been on historical abuse – events that happened in the "dim and distant past". In many cases the perpetrators were deceased or no longer well enough to face a court trial. It was time to draw a line and move on. Things like that wouldn't happen nowadays!

Were people right to assume that the nightmare had come to an end? Are there no more dirty little secrets? A glance at one of our local newspapers in Northern Ireland, on a date picked at random, provides the answer.

On July 31, 2019, it was reported that a final year computer science student had appeared in a Magistrates' Court following what is thought to be the UK's biggest internet child abuse investigation. The charges involved 45,000 child abuse images. The court heard that the defendant would befriend a child by pretending to be someone else before asking them for an image. This is known as 'catfishing' – creating a fictional online persona to lure children, or vulnerable adults, into the net. He would later write to the alleged victim telling them if they didn't do as he said, he "would show your nudes for all the world to see".

More than 300 alleged victims had been identified. The prosecution said that the children targeted, aged between 10 and 12 and mostly female, were "left in distressed states". The investigators found 'Snapchat' maps on seized devices locating the children. These are thought to have been for future reference.

The images under investigation were for sale on fraudulent female 'PayPal' accounts. Prices ranged from $20 and $50 for 'CA' – thought to mean 'Child Abuse' – material. The defendant told police that he needed the money to pay off gambling debts.

The court heard that following his arrest and release on police bail, the defendant continued to Snapchat other young people the very next day. The District Judge adjourned the case for four weeks and remanded the defendant in custody.

A separate article in the same newspaper reported that a man caught with 3,000 images of girls on his mobile phone and laptop was placed on probation for 30 months.

A third article reported that a convicted sex offender had been found, dressed as a woman, in a school. He was jailed for four months.

The IICSA's report on Child Sexual Abuse in the Church of England, mentioned above, is based on the inquiry's public hearings held in July 2019. The report states that, in 2018, the latest date for which figures were available, there were 2,504 safeguarding concerns reported to dioceses in England about either children or vulnerable adults. 449 of these were about recent child sexual abuse. A significant number of the 449 involved the downloading or possession of indecent images of children.

Clearly, child sexual abuse has not gone away. Abuse, and the secrecy around it, remains a problem. The true gravity of the phenomenon remains unknown. What has changed is the nature of the abuse. With the expansion of the Internet, the growth of social media and the increasing sophistication of mobile phones, the sexual exploitation of children has become a form of cybercrime. In 2017, an INTERPOL report led to the identification of 14,289 victims in 54 European countries. Consequently, the sexual exploitation of minors is listed as one of the nine EMPACT priorities, Europol's priority crime areas, under the 2018-2021 EU Policy Cycle.

The Internet Watch Foundation (IWF) is a charity that searches for and removes online child sexual abuse imagery. Their Annual Report 2017 shows online child sexual abuse imagery up by 37% on 2016. 78,589 URL[49]s were identified that contained images of sexual abuse, concentrated particularly in the Netherlands, followed by the United States, Canada, France and Russia. Overall, 87% of all child sexual abuse URLs identified globally in 2017 were hosted in just these top five countries.

---

[49] A URL – Uniform Resource Locator – is a reference to a web resource that specifies its location on a computer network and a mechanism for retrieving it. It is effectively a web address. In the current context the URL identifies the location of child sexual abuse imagery.

The Report makes for uncomfortable reading. 55% of the victims are less than 10 years old. 86% contained images of girls, 7% of boys, and 5% contained images of both boys and girls. The images and videos found have increased in their severity. The most serious Category A images, depicting rape and sexual torture, rose to 33% from 28%. Category B images rose from 19% to 21%.

The IWF also found an 86% rise in the use of disguised websites, from 1,572 in 2016 to 2,909 in 2017. These are websites where the child sexual abuse content is only revealed to someone who has followed a pre-set digital pathway. To anyone else, they will only show legal content. It's concerning that offenders appear to be increasingly using concealed digital pathways to prevent law enforcement and hotlines around the world detecting these criminal websites[50].

Deborah Denis, Head of Fundraising and External Relations at The Lucy Faithfull Foundation, said:

"The shocking figures (in IWF's Annual Report) are a timely reminder that the threat to children posed by illegal online child abuse material is very real and is growing all the time. Put simply, behind every image counted in this report is a child experiencing the unimaginable trauma that comes with the knowledge that an image of their abuse is being viewed and shared across the internet."

NSPCC Associate Head of Child Safety Online, Andy Burrows said:

"It's clear that paedophiles are using increasingly sophisticated ways to offend on a mass scale. The use of disguised websites and the dark web are fuelling the growth of this terrible crime... The sheer scale and complexity of the problem is evolving rapidly in line with technology, so it's impossible to simply police our way out of the problem, we need a comprehensive strategy to stop potential offenders in their tracks. We know a lot of child sexual abuse imagery is created after predators have groomed their victims and to tackle the growth of this material, we need to cut it off at the source. Social networks must ensure that they prioritise child protection."

Susie Hargreaves OBE, IWF CEO, says: "The child victims of sexual abuse online are re-victimised again and again, every time their picture is shared. The experience they go through at such a young age is unimaginably horrific, and they frequently take this pain into adulthood with them. ...There is a vast amount of content out there. Sadly, this could just be the tip of the iceberg. We are making huge technological advances, which we'll be announcing later in the

---

[50] To read the IWF's full 2017 Annual Report go to: https://annualreport.iwf.org.uk/

year, but we also need to continue to work globally, in partnership, to fight this disturbing crime. This battle cannot be won in isolation."

There is one group of offenders I have not mentioned as yet. These people would claim to be tourists, just like us. But they are not just tourists, they are paedophiles whose insatiable appetite for underage sex overrides any concerns for the welfare of children.

The perpetrators of such abuse, in most cases, claim to be unaware of the fact that they are committing a crime. It is difficult to accept that someone who is fully aware that underage sex is a criminal offence back home could believe that it is perfectly acceptable in far foreign fields.

2017 data released by the World Tourism Organization (UNWTO), shows that three million persons each year take a trip in order to have sexual relations with minors. The most popular destinations are Brazil, the Dominican Republic, Colombia, Thailand and Cambodia, and more recently, some African and Eastern European countries.

The first six countries of origin of those who perpetrate the abuse are: France, Germany, the United Kingdom, China, Japan and Italy. Not to be overlooked is the growing number of women traveling to developing countries seeking paid sex with minors. They represent about 10% of the world's sex tourists.

According to a study conducted by ECPAT International (End Child Prostitution in Asian Tourism) between 2015 and 2016, 35% of these sex tourists are regular customers, while 65% are occasional customers.

It makes little difference to the victim whether the person who comes in to defile them is a serial offender or a first timer. The sexual abuse of children, in whatever form, is a crime and must be stopped.

We already know a great deal about what must be done to eliminate sexual trafficking and the abuse of children. Using the force of the law to prosecute perpetrators must be a key goal. Extraterritorial legislation is required in order to prosecute nationals who have committed offences against children in other countries. Although addressing only one aspect of a larger problem, these laws must be used aggressively to bring an immediate halt to trafficking and profiting from exploited children.

Child exploitation is a multi-billion-dollar, multi-national business. It can only be dealt with at an international level. It is also multi-faceted. There is no one solution but many, each tailored to the diverse national, local and cultural realities in which these affronts to child rights originate.

To succeed, we must strengthen international cooperation and action at every level of every society. Governments alone will not succeed. Partnerships of governments, intergovernmental and nongovernmental organisations are needed to be really effective.

It is up to all of us – governments, law enforcement, international organisations and all levels of civil society, to see to it that the elimination of commercial sexual exploitation is accorded urgent priority.

We know that it is often the very adults entrusted with the care and protection of children who sexually exploit them. But whether the offender is an individual or part of a criminal gang there must be no more "turning a blind eye".

Governments and media outlets must shine a light in the dark corners where exploitation and abuse occur so that the abusers are exposed for all to see. This might entail public information campaigns, increased media coverage, more sophisticated monitoring and sharing of information, educating children about sexual abuse from an early age at home and in school.

It may appear at first sight that there is no specific role for members of the IAYFJM. As a general rule the role of judges and magistrates is *after* the event, dealing with victims and perpetrators. But, in doing that, they can play a crucial role.

Some of the most important improvements in support of child victims who must testify in court do not require legislation, but rather an educated and assertive judiciary. Judges can control the process of examination and cross-examination of children. They can, and should, put a stop to children being intimidated, intentionally confused, or harshly spoken to by defence lawyers. They can insist that questions lawyers seek to ask young children be submitted to the judge in advance, and then the judge can edit these to check linguistic appropriateness and make sure the questions are short and unambiguous.

Where possible, abused children should be allowed to have their evidence recorded on video and presented at the trial as "Evidence in Chief". Any necessary cross-examination should be done by video link so that the child need never face the alleged offender in court.

In countries where legislation does not permit this approach, judges can modify the courtroom setting, including the location of the judge, the defendant, and the child witness to reduce the child's stress.

Child witnesses should also be allowed to visit the courtroom before the hearing. Where possible, 'court school' programmes should be arranged to prepare them for the trial experience, and to provide them with colouring books or videos that help explain the process.

Judges should be authorised to appoint a legal representative for the child victim in criminal court – a 'guardian ad litem' who would be responsible for protecting the child's rights and interests in connection with testifying and other purposes. This person could help assure the child is not re-victimised in the judicial process.

Judges should be authorised to permit a child's testimony via international video-link if the child is located in a country other than where the offender is being prosecuted.

Judges might call for harsher penalties for those found guilty of the sexual exploitation of children.

Convicted sex tourism offenders should be subject to forfeitures (with confiscated funds/property benefiting prostituted children and other child victims of sexual exploitation) as well as made responsible for restitution to the child victim in his or her country of residence.

Judges should be aware of international treaties that can be applied, (although they will clearly only be applicable if the two nations concerned have ratified them!) These are:

The Hague Convention on the Civil Aspects of International Child Abduction;

The Hague Convention on Protection of Children and Co-Operation in Respect of Intercountry Adoption; and

The Hague Convention on Jurisdiction, Applicable Law, Recognition, Enforcement and Co-Operation in Respect of Parental Responsibility and Measures for the Protection of Children.

The IAYFJM could work towards mobilising members of the legal profession – lawyers and judges – to secure the rights and protect the interests of child victims of sexual exploitation. A database of legal advocates with a particular interest in helping children involved in prostitution, pornography, sexual trafficking, cross-national internet crimes and other related offences might be drawn up.

Covid-19 is likely to have a major impact on how we do things. It seems inevitable that there will be less travelling and greater use of the Internet. The IAYFJM recently launched a series of monthly Webinars to replace international conferences. At 2:30 GMT on July 22, 2020, for example, Professor Ann

Skelton, University of Pretoria, South Africa, a member of the UN Committee on the Rights of the Child, conducted a Webinar on the Committee's General Comment No 24. Professor Skelton introduced the document. Spanish Judge Dr Jorge Jimenez, Director of the Spanish Judicial School and Director of the Ibero-American Network of Judicial Schools, addressed the challenges for training. Justice Imman Ali, Justice of the Supreme Court of Bangladesh, Appellate Division, focused on the implementation. Professor Skelton was in her office in South Africa, Judge Jimenez was in Spain and Justice Imman Ali in Bangladesh. The audience consisted of all those IAYFJM members who had registered for the event. On registration, members were given the web address which went "live" at the time scheduled. This enabled them to participate in the Webinar from wherever in the world they happened to be. Amazing!

The webinar was posted in the Association's YouTube page where members who missed the live webinar could have access to it (and all previous webinars), with the possibility of having subtitles in many languages. These "Webcasts" (webinars recorded for broadcasting) are not time-constrained so they can be viewed and reviewed at any time, day or night.

It seems likely that this approach will become the norm in the future. Members of the Executive Committee have, in the recent past, been able to participate in meetings remotely when unable to travel. But poor connections did not encourage general use. Advanced technology means that Zoom meetings are fully interactive and allow all participants the ability to see, speak, hear, and screenshare with each other. This is ideal for committee meetings or for working groups where members can share relevant legal articles, laws and court decisions or discuss common transnational law reform strategies.

Webinars, such as the one mentioned above, have been described as "interactive seminars conducted over the internet". But they are only interactive to a certain extent. A webinar may contain audience polls, Q&A chat functions, and whiteboards. Attendees can type in and submit questions to the speaker during the live session. But the speaker won't be able to see or hear the audience while presenting a webinar and the attendees won't be able to see or hear one another.

It is to be expected that the agencies I have worked with over the years – UNICEF, UNDP and the Council of Europe – will make use of webinars in their training programmes. This will increasingly be the case as webinars become fully interactive. However, no matter how good the Internet presentation, nothing can equal a face-to-face presentation. This is particularly the case with developing nations since attendees will have had little experience of training via the Internet. Judges and lawyers working in oppressive regimes will be reluctant to discuss contentious issues knowing that the webinar will be viewed and reviewed by the authorities later. I have had judges who would not speak out in the presence of their colleagues but who would confide in me when I worked with them on a one-to-one basis.

International missions by experts have proven to be eminently successful in addressing child abuse and inequality and in providing an educated and assertive judiciary. While accepting the inevitability of travel restrictions as a result of Covid-19, I am firmly of the view that these missions must continue. With that in mind, our members should hold themselves ready and willing to assist in the training of judges, lawyers and court personnel, both nationally and internationally.

Conclusion

Travel allows us to see the world in ways we would never be able to understand sitting at home and reading about it. Standing on the bridge over the river Kwai and walking through the streets of Warsaw, thinking about the horrors of war and man's inhumanity to man, impacted much more intensely on me than sitting at home, in the comfort of an armchair, book in hand, or watching a film in the cinema.

No book or film could convey the feelings aroused on coming round a corner in a South African safari park, to find a giant bull elephant blocking the road. Towering above us in our tiny jeep, his trunk swinging slowly from side to side, he looked like he was spoiling for a fight. He stared at us intently as if to say: "Come on, try to pass, make my day".

While I enjoy visiting exotic locations, seeing beautiful scenery and having an opportunity to study the flora and the fauna, what I look forward to most is interaction with people. In my view, it is possible to relate to anyone in the world if you look past the superficial things that separate you. A smile, and a friendly attitude, can break down barriers and help create friendships.

I got an opportunity to test my theory on my first visit to Beijing. I had no experience of Chinese police officers. What I knew about them was based on media reports of the *Chengguan*, or "city management officials". These are a particular type of law-enforcement officers who enforce municipal regulations in Chinese cities. Through a seemingly never-ending series of high-profile incidents, *Chengguan* captured the public attention and attained a remarkable degree of notoriety in China. They have a reputation for being brutal, unnecessarily violent and unpleasant. I did not know if this reputation, which, I assumed, applied to all police officers, was fair.

My hosts provided me with a police car and driver for a day to take me wherever I wanted to go. This was my first contact with any police officer on a one-to-one basis in China. I didn't know what to expect. I followed my own guidelines – he was a friend I was just getting to know. We built up a good relationship throughout the day. (see p81)

But it takes two to tango and sometimes the other party has difficulty seeing beyond the superficial things that separate the two of you. Their life experience may have taught them to be suspicious of strangers who make friendly advances. I have frequently found that people saw me as representing the UN or the US or both and this made them guarded in their approach. It is important to try to understand where they are coming from and that it may take some time for them to accept that you are genuine.

Wei Long, a Chinese Supreme Court judge, spent three hours in Sion, Switzerland trying to convince me that the assault on protesters in Tiananmen Square (in 1989) had not in fact happened. Over the years, he learned that I was not prejudiced against the Chinese. I did not judge people by the colour of their skin, their religion or their ethnicity. I did not hold individuals responsible for the actions of others. We became firm friends. As we sat in his apartment in Beijing, years after we first met in Sion, he told me what it was really like. For good measure, he told me about his personal experiences of the *Cultural Revolution*.

Xui Li, Wei Long's successor as Chinese representative on the Executive Committee of the IAYFJM, had a very stern exterior and seldom smiled in public. She was very much aware that, in her role as a member of the Committee, she represented the Chinese State and was answerable to the Communist Party. However, it was clear to me that she wanted to be seen as an individual, someone with her own identity, and not just as the representative of the Chinese Government. When I nominated her for election as an Honorary Member of our Association, because of her efforts to reform her country's policy on the death penalty, she was like a child for whom Christmas had come early. She was literally jumping with joy, not knowing whether to laugh or cry. She was so pleased that her efforts were being validated. *She* was being recognised and not just the Communist Party.

It was clear that Valentina Semenko, the Russian Supreme Court judge, felt the same way. She broke down and cried when her ambassador refused her permission to come with us to Beijing. She just wanted to relax for a few days with her friends. The Ambassador said "No!" She was so different when I met her in Ivanovo. I had responded to *her* request to lend my support, as President of the IAYFJM, to the setting up of a pilot Juvenile Court. I had validated her efforts to improve the lot of children in Russia. I had acknowledged her as my friend and had come to help her out. I had never seen her smile at Executive Committee meetings. She was smiling now.

While working in Iran in 2006, I found the judges and prosecutors somewhat "distant" at first. I guessed they were looking at me as a representative of the UN and remembering the sanctions. I was discussing the programme with them and suggested times for coffee breaks, lunch etc. The judges were very quick to tell me that they would be stopping for prayer at fixed times. It was clear from their tone that they were telling me – not asking for permission. I told them I fully supported their wish. I explained that we had a custom in Ireland where the ringing of the Angelus bell called people to prayer. In the past people would stop whatever they were doing and recite the Angelus. I said I was sorry to say that the custom was dying out. Many churches no longer ring the bell to signal the Angelus and, where they do, few people stop to pray. I told the judges that I hoped they would not abandon their calls to prayer. Once the group realised that I respected their customs all resistance disappeared and we built up a good rapport. We were able to discuss the thorniest of issues.

It's not always possible to make that kind of breakthrough. There may be insufficient time to build up a relationship. Or it may be that the other person doesn't want one. In my report from Argentina, I comment on the chair of the local organising committee for the IAYFJM's World Congress in Buenos Aires in 1998. A big, domineering man, he would accept no advice and brooked no questioning of how he organised things. He wasn't interested in making friends. He just wanted people to keep out of his way.

The judges in Kosovo had different reasons for not wanting to be friends (see p135). They were resentful at being forced to do refresher courses. They were particularly resentful at being made to study the European Convention on Human Rights (ECHR). One judge told me that I didn't understand and went on to outline an example of a Serb commander ripping a child from its mother's arms, decapitating it and then handing the headless baby back to its mother. Should they respect his human rights?

Where our views on a country and its peoples are based on a single visit those views may be biased according to whether our experience was good or bad. A little knowledge can lead to misconceptions. Repeated visits or a period of residence can provide a more balanced view.

Before visiting Myanmar, I had a very positive image of Buddhism. This was reinforced during my visits when I learned that the village monastery functions as the centre of social activities, where most communal affairs take place.

This positive image took a knock on the train from Bangkok to Kanchanaburi when I witnessed a Buddhist Monk throw the rubbish out the window when he finished his lunch. So much for protecting the environment.

The image took a more serious knock when I became aware of a monk called Ashin Wirathu who presided over a very large monastery of some 2,500 monks, preaching hatred of Muslims and encouraging Buddhists to drive them out of Myanmar. Initially, this appeared to be an extremist minority. Later it became apparent that Myanmar's military has a policy of ethnic cleansing and has been accused of committing atrocities in efforts to achieve its aims.

In trying to find out what lay behind the hatred of Muslims I discovered that tensions between the Buddhist majority and the Rohingya Muslims date back to the beginning of British rule in 1824. As part of their divide-and-rule policy, British colonists favoured Muslims at the expense of other groups, causing deep resentment amongst the Buddhist majority. While this doesn't excuse the ethnic cleansing, knowledge of the deep roots is a necessary prerequisite to understanding what is going on.

This ebb and flow of positive and negative information about Myanmar and the Burmese[51] over a series of visits allowed me to get a more balanced view of what life is like there than the one I started with.

Experiencing new cultures, seeing how other people live their lives, enriches us. We see the world in a whole new light. Our mind grows and expands irrespective of whether the experiences are good or bad.

My conclusion, after a lifetime of travelling, is that people are the same the world over.

I have been impressed by the faith shown by people of all religions in the many countries I have visited. True there are some, like Ashin Wirathu in Myanmar, for whom a profession of deep faith in their God is followed by a tirade of hatred against their enemies. However, the vast majority of ordinary people, whether they be Buddhists with their prayer wheels, Muslims with their prayer beads or Catholics with their rosary beads, pray to their God for a better future for themselves and for their children.

---

[51] The generals didn't change the official name of the people when they changed the name of the country.

We all have the same basic wants and needs. It doesn't matter where we come from, what we look like, how we talk, or what we believe. Deep down the essence of who we are as human beings is the same. Whether the skin is black, brown or white, the love in a mother's eyes is the same everywhere. Children laugh and play and clap their hands in Monrovia just as they do in Yangon, Cape Town, Melbourne, Montreal, Rio de Janeiro or Belfast.

People hug each other at the arrival/departure gate of Moscow's Domodedovo airport in the same way as they do in Belfast International. Love speaks the same language, wherever we are. We *all* have a dream – a dream of love, security, enjoyment and hope for a better future.

Whether your dream is to travel the world, learn more about your country of birth or adoption, explore your parish or research the past – what are you waiting for? Don't put it off until tomorrow. Set your travel dreams in motion today.

"Twenty years from now you will be more disappointed by the things that you didn't do than by the ones you did do. So, throw off the bowlines. Sail away from the safe harbour. Catch the trade winds in your sails. Explore. Dream. Discover."

Mark Twain.

Countries I Have Been To

1	Andorra	26	Greece	51	Paraguay
2	Argentina	27	Holland	52	Peru
3	Australia	28	Hong Kong	53	Philippines
4	Austria	29	Hungary	54	Poland
5	Bahamas	30	India	55	Russia
6	Belarus	31	Iran	56	San Marino
7	Belgium	32	Ireland	57	Scotland
8	Botswana	33	Israel	58	Singapore
9	Brazil	34	Italy	59	Slovakia
10	Canada	35	Japan	60	Slovenia
11	Chile	36	Jordan	61	South Africa
12	China	37	Kenya	62	Spain
13	Colombia	38	Kosovo	63	Sweden
14	Cote D'Ivoire	39	Liberia	64	Switzerland
15	Cuba	40	Lithuania	65	Tajikistan
16	Cyprus	41	Luxembourg	66	Tanzania
17	Czech Republic	42	Malaysia	67	Thailand
18	Denmark	43	Mexico	68	Tunisia
19	Ecuador	44	Monaco	69	Turkey
20	Egypt	45	Myanmar	70	Turkmenistan
21	England	46	Nepal	71	Uruguay
22	Finland	47	New Zealand	72	USA
23	France	48	Norway	73	Vatican
24	Georgia	49	Palestine	74	Venezuela
25	Germany	50	Panama	75	Wales

Timeline

1961 Qualified as a teacher.

1974 Appointed University Lecturer (Teacher Training).

1976 Appointed a Lay Magistrate (Belfast Youth Court and Belfast Family Proceedings Court).

1982 Elected Honorary Secretary of the Northern Ireland Lay Magistrates' Association (NILMA).

1982 Appointed Northern Ireland representative on the Executive Committee of the British Juvenile & Family Courts Society (BJFCS).

1986 Invited to represent Northern Ireland at the World Congress of the International Association of Youth & Family Judges & Magistrates (IAYFJM) in Rio de Janeiro, Brazil.

1990 Elected Chair of BJFCS.

1990 Elected onto the General Purposes Committee of the IAYFJM at the World Congress in Turin.

1991 Appointed Editor-in-Chief of the IAYFJM's magazine.

1992 Elected Chair of NILMA.

1994 Elected onto the Executive Committee of the IAYFJM at the World Congress in Bremen.

1998 Elected Vice President of the IAYFJM at the World Congress in Buenos Aires.

2002 Elected President of the IAYFJM at the World Congress in Melbourne, Australia.